Cross-Sectional Atlas of Human Brainstem

Jin Seo Park • Yaqian You

Cross-Sectional Atlas of Human Brainstem

With 0.06-mm Pixel Size Color Images

Springer

Jin Seo Park
Department of Anatomy
Dongguk University School of Medicine
Gyeongju-si, Gyeongsangbuk-do
Korea (Republic of)

Yaqian You
Department of Anatomy
Dongguk University School of Medicine
Gyeongju-si, Gyeongsangbuk-do
Korea (Republic of)

ISBN 978-981-19-9521-7 ISBN 978-981-19-9519-4 (eBook)
https://doi.org/10.1007/978-981-19-9519-4

This Springer imprint is published by the registered company Springer Nature Singapore Pte Ltd.
The registered company address is: 152 Beach Road, #21-01/04 Gateway East, Singapore 189721, Singapore

Contents

1 Introduction

1.1 Introduction

The brainstem is the neural center relaying afferent and efferent neural information between the cerebrum with the cerebellum and the whole body through the spinal cord. Since a brainstem consists of numerous organized nuclei and fibers in the narrow and compact space, a fine map of brainstem is required for neurosurgeons and neuroscientists who should know neuroanatomy.

Existing brainstem atlases were made from histological stained images, magnetic resonance images, or schematic figs [1, 2], whose images could not be colored or either gray matter or white matter could be shown. Therefore, to make a fine map of the brainstem, we needed new images of the brainstem that both the nucleus and the fibers can be displayed in true color and high resolution. By this necessity, we choose the true color sectioned images of Visible Korean [3, 4].

We have already published the *Cross-Sectional Atlas of the Human Head* in 2018 [5] from head sectioned images in 2009 [3]. In 2021, we made new sectioned images of the male whole body again [4]. Based on the *Cross-Sectional Atlas of the Human Head* publishing experience, this cross-sectional atlas of the human brainstem was made from the new sectioned images.

J. S. Park, Y. You, *Cross-Sectional Atlas of Human Brainstem*, https://doi.org/10.1007/978-981-19-9519-4_1

1.2 Making the Sectioned Images and Atlas of the Brainstem

In 2005, we have produced the sectioned images of the male whole body with 24-bit color and 0.2 mm × 0.2 mm sized pixel, which is the first edition of Visible Korean [6]. However, the brainstem could not be observed in detail in that dataset because of the herniation of the brainstem. Therefore, in 2021, we produced new axial sectioned images of the male whole body. Intervals, pixel size, resolution, and color depth of the images were 0.2 mm (head), 0.06 mm × 0.06 mm, 8688 × 5792, and 48-bit color, respectively [4]. From the sectioned images of the whole body, the images including only the brainstem were selected. By the computer-assisted three-dimensional reconstruction, coronal and sagittal sectioned images (intervals, 0.06 mm; pixel size, 0.06 mm × 0.2 mm) were produced from the axial images (intervals, 0.2 mm, pixel size, 0.06 mm × 0.06 mm) (Fig. 1.1).

In the axial, coronal, and sagittal sectioned images, 154 structures including nuclei, tracts, and other structures were identified and annotated by referring existing atlases and anatomy books [1, 2, 5]. Some pathways of tracts were marked in the images using dotted line arrow. To find and trace the structures accurately, we used the volume models from brainstem sectioned images. In the volume models, not only sectioned images of three planes but also three-dimensional models could be displayed in real time (Fig. 1.1).

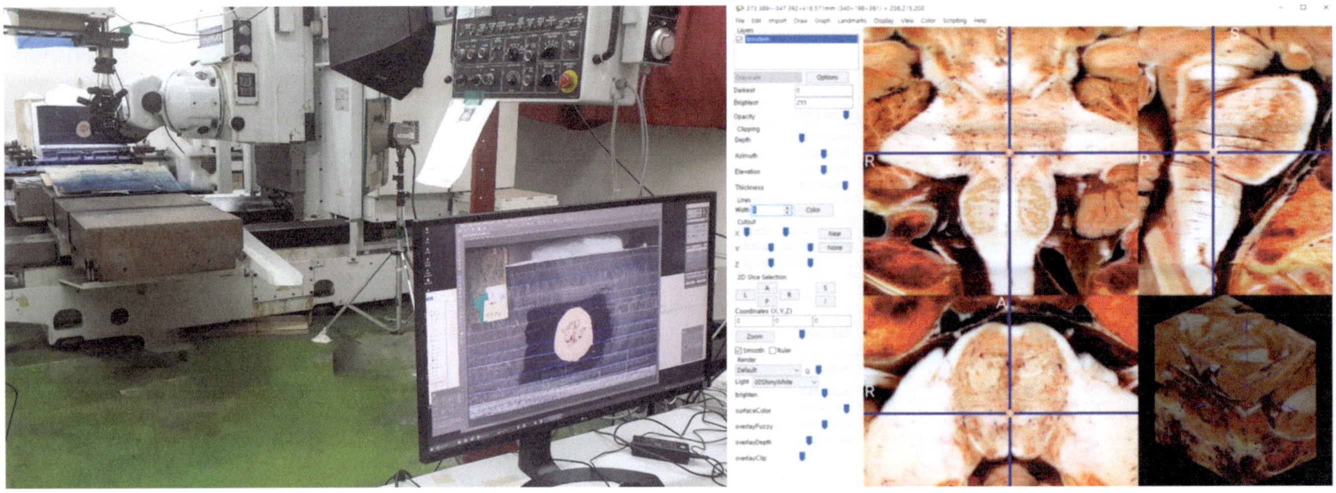

Fig. 1.1 Sectioning the cadaver, and photographing the sectioned surface to make a sectioned image (left). Finding and tracing the structures of the brainstem in axial, coronal, and sagittal sectioned images in real time in three-dimensional volume model using MRIcroGL software (right)

1.3 Terminology

Accurate anatomical terms are essential to the anatomy atlas. Therefore, all anatomical terms in this atlas were labeled based on the authorized terms, Terminologia Anatomica [7].

Acknowledgments This work was fully supported by Anatomage Inc., USA. We are deeply grateful for this assistance. This research was also supported by a Basic Science Research Program through the National Research Foundation of Korea (NRF) funded by the Ministry of Education (NRF-021R1F1A1063044).

If you find something wrong in this atlas, please let me know (Jin Seo Park, park93@dongguk.ac.kr, neuroanatomy.kr).

References

1. Naidich TP, et al. Duvernoy's atlas of the human brain stem and cerebellum. 1st ed. Germany: Springer; 2008.
2. DeArmond SJ, et al. Structure of the human brain: a photographic atlas. 3rd ed. New York: Oxford University Press; 1989.
3. Park JS, et al. A proposal of new reference system for the standard axial, sagittal, coronal planes of brain based on the serially-sectioned images. J Korean Med Sci. 2010;25:135–41.
4. You Y, Park JS, et al. Advanced-sectioned images obtained by microsectioning of the entire male body. Clin Anat. 2022;35:79–86.
5. Park JS. Cross-sectional atlas of the human head : with 0.1-mm pixel size color images. Germany: Springer; 2018.
6. Park JS, et al. Visible Korean human: improved serially sectioned images of the entire body. IEEE Trans Med Imaging. 2005;24:352–60.
7. Drake RL. Terminologia Anatomica: international anatomical terminology by FIPAT. 2nd ed. Stuttgart: Thieme; 2011.

Axial Brainstem

2

Images of −1.0 mm ~ −2.4 mm, Diencephalon and midbrain

Images of −3.0 mm ~ −12.4 mm, Midbrain

Images of −13.0 mm ~ −14.4 mm, Midbrain and pons

Images of −15.0 mm ~ −32.4 mm, Pons

Images of −33.0 mm ~ −36.4 mm, Pons and medulla oblongata

Images of −37.0 mm ~ −57.4 mm, Medulla oblongata

Images of −58.0 mm, Spinal cord

All images, 0.06 mm × 0.06 mm sized-pixel and 0.4 mm or 0.6 mm intervals

Abbreviation: a., artery; ant., anterior; inf., inferior; int., internal; lat., lateral; m., muscle; med., medial; n., nerve; nu., nucleus; post., posterior; sup., superior; tr., tract

J. S. Park, Y. You, *Cross-Sectional Atlas of Human Brainstem*, https://doi.org/10.1007/978-981-19-9519-4_2

Diencephalon and midbrain (Axial, -1.0 mm)

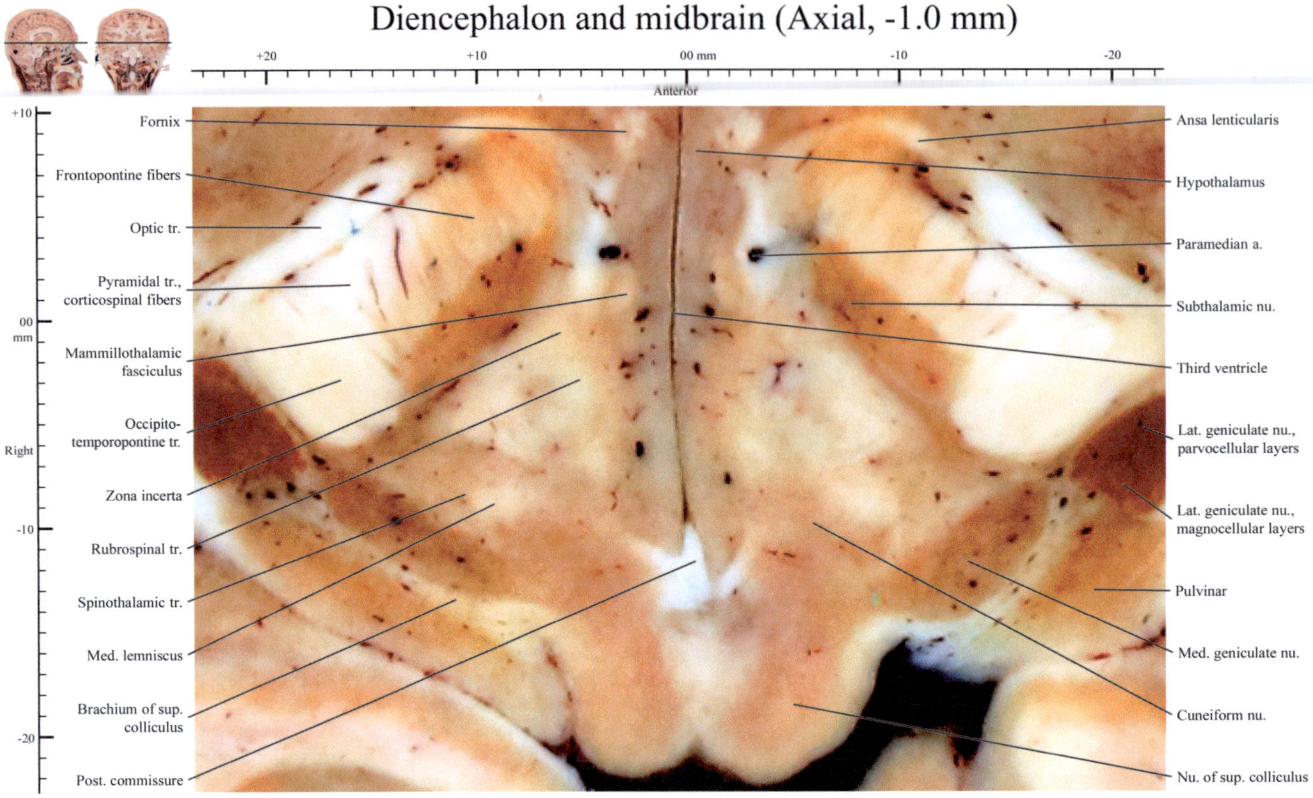

Diencephalon and midbrain (Axial, -1.4 mm)

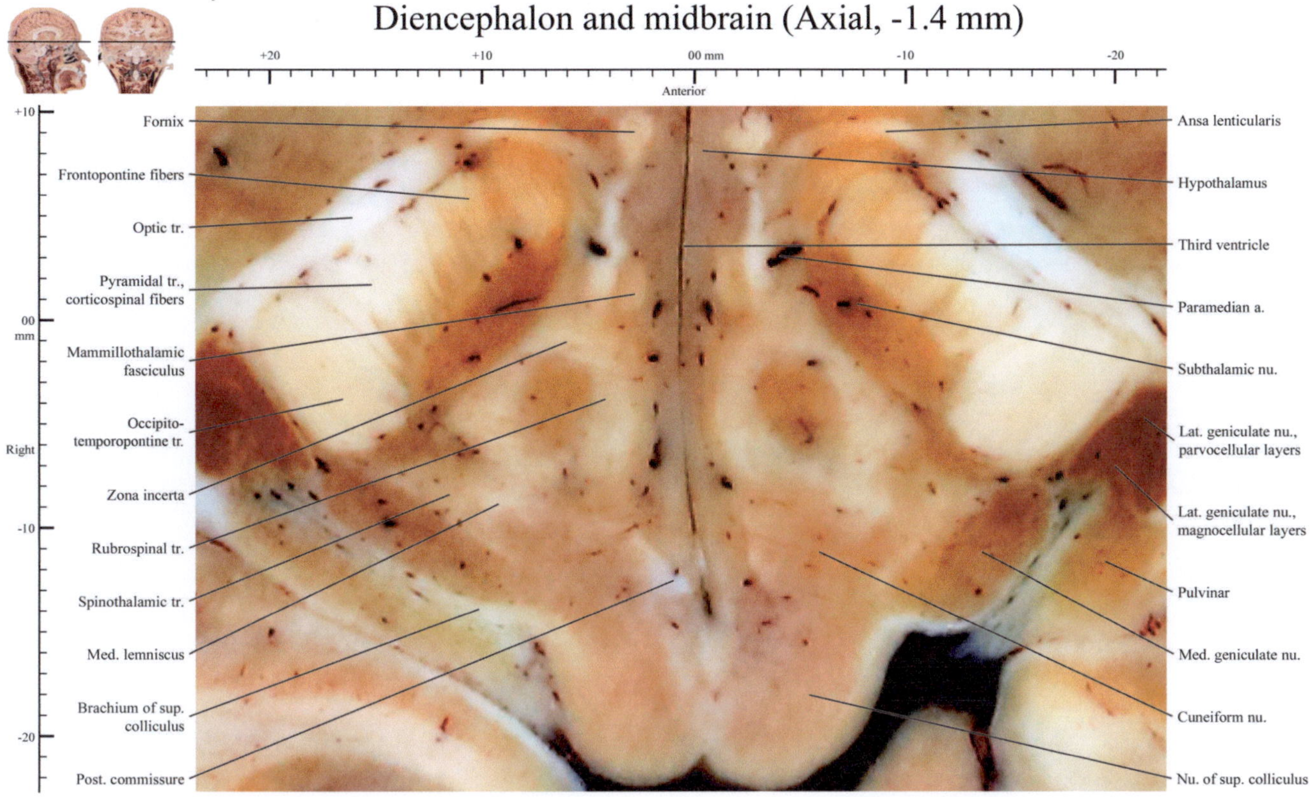

Diencephalon and midbrain (Axial, -2.0 mm)

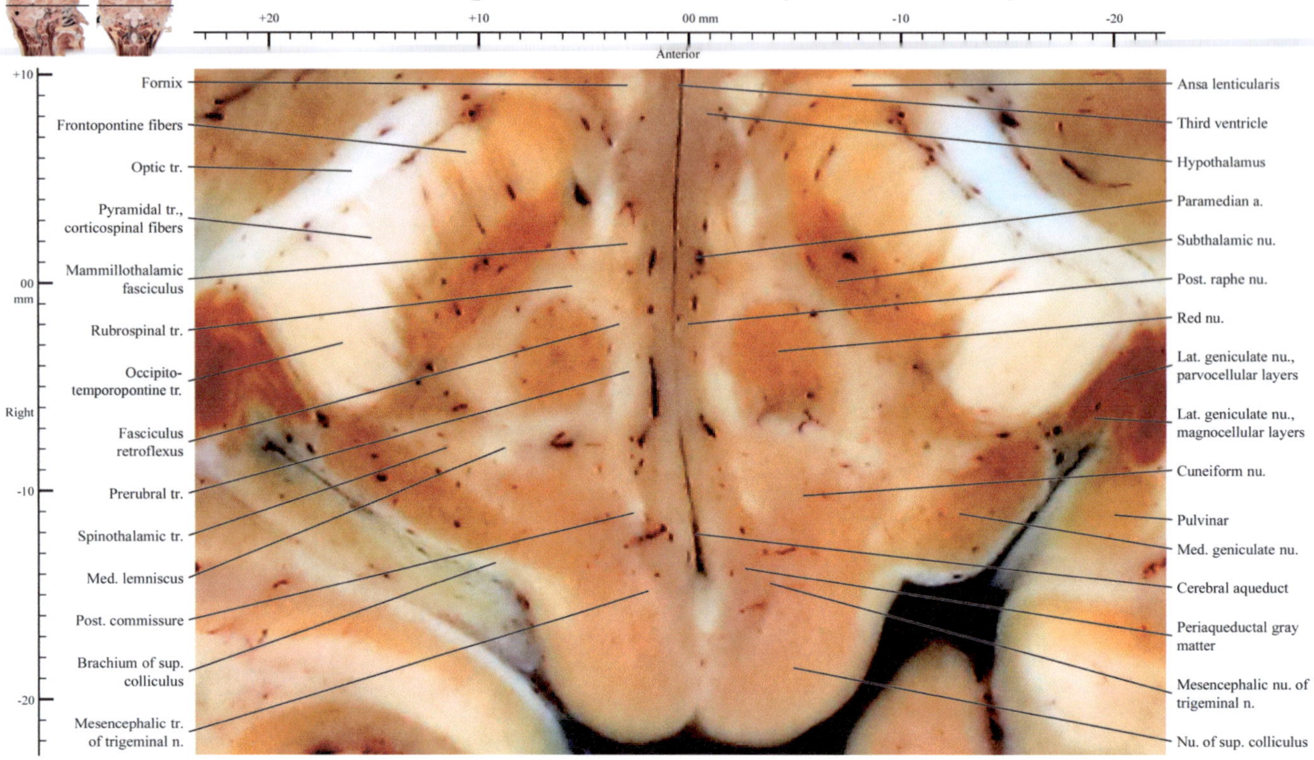

Diencephalon and midbrain (Axial, -2.4 mm)

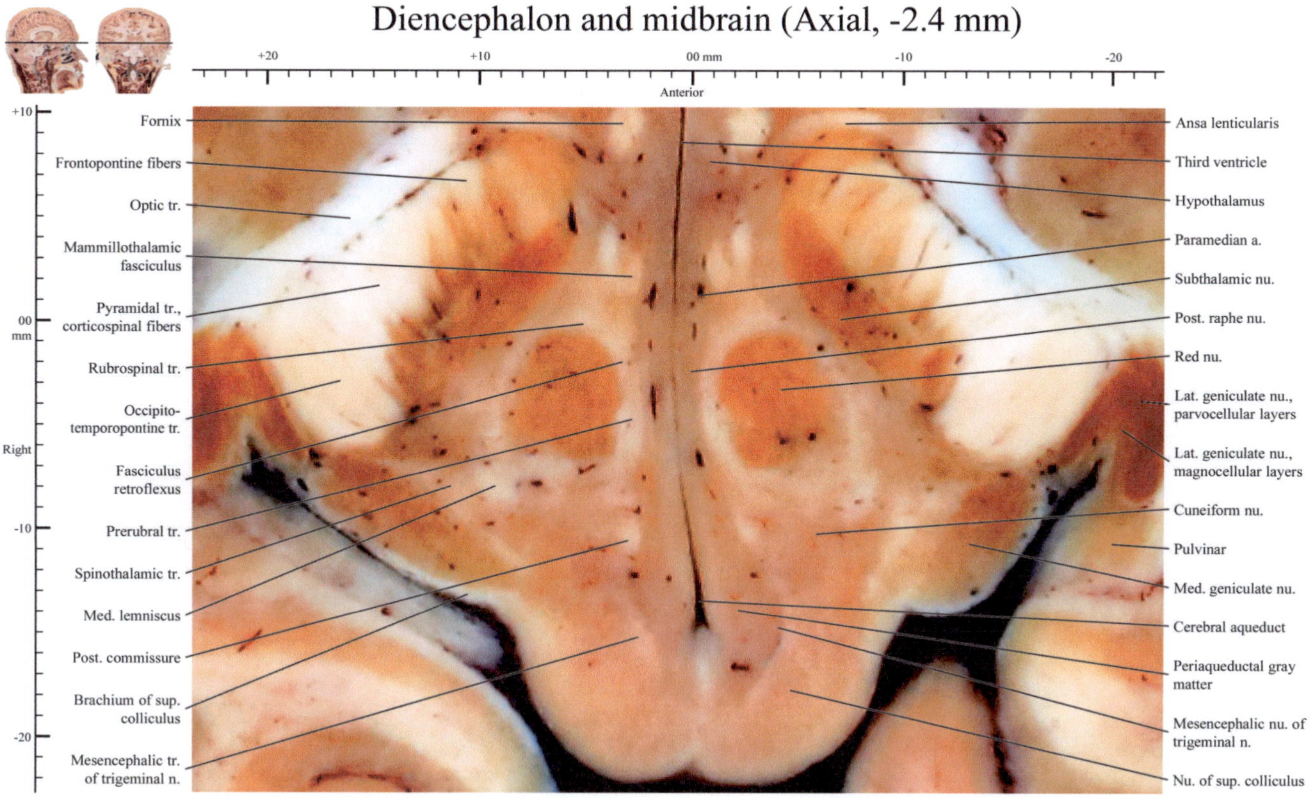

Midbrain (Axial, -3.0 mm)

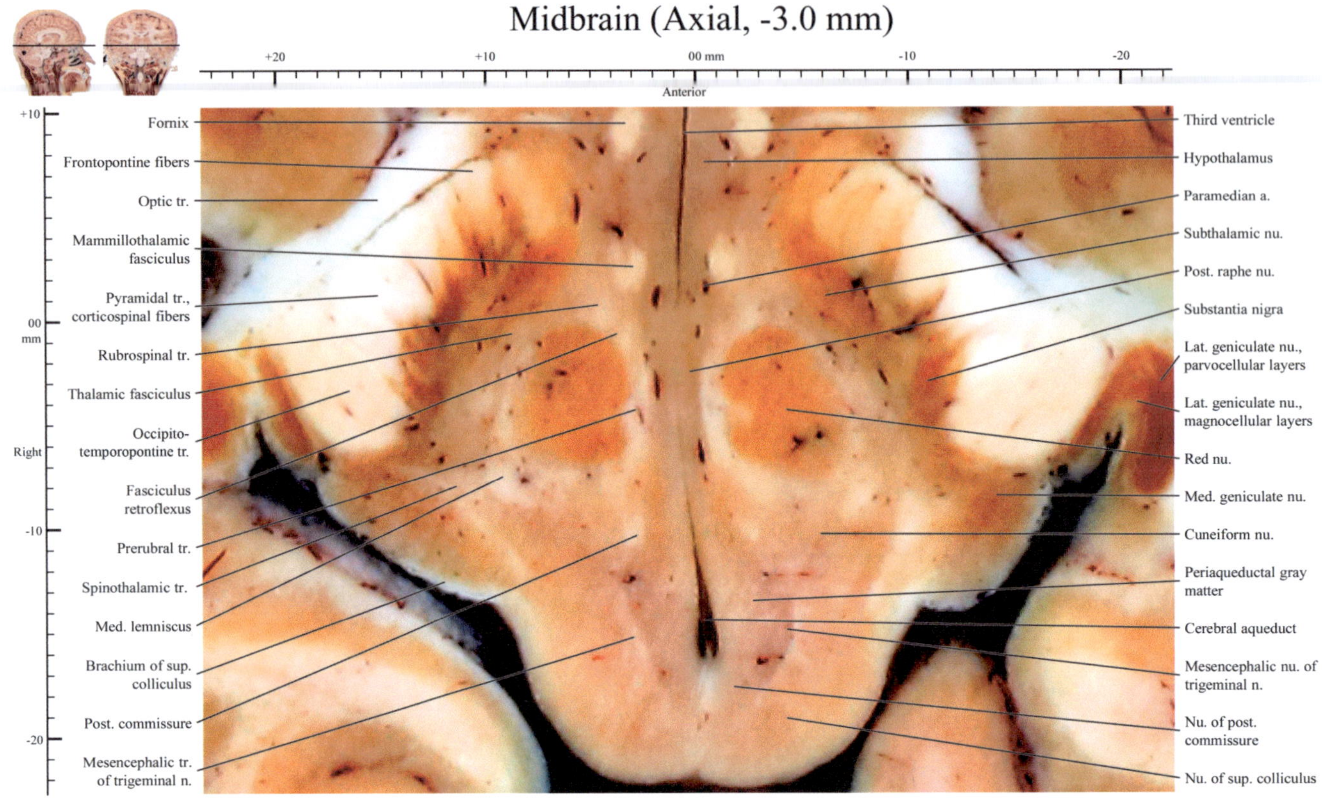

Midbrain (Axial, -3.4 mm)

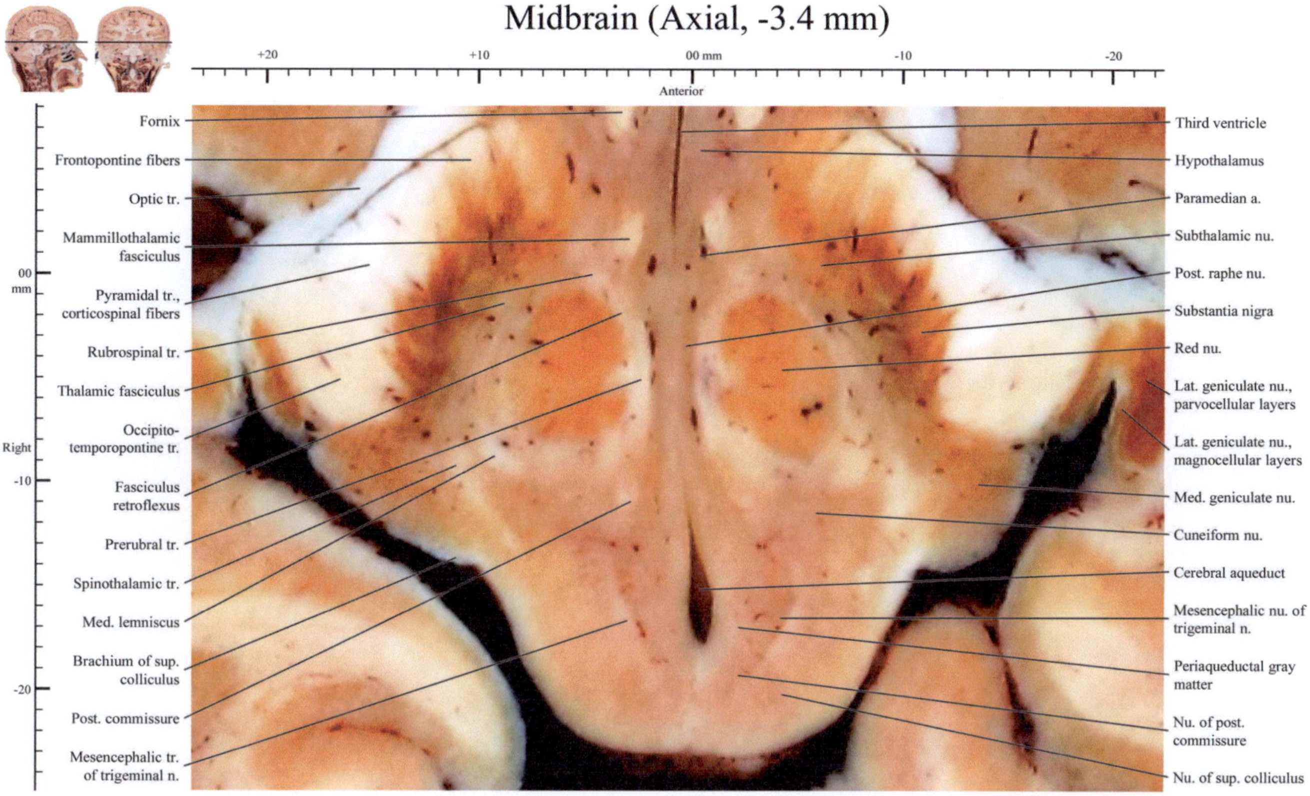

Midbrain (Axial, -4.0 mm)

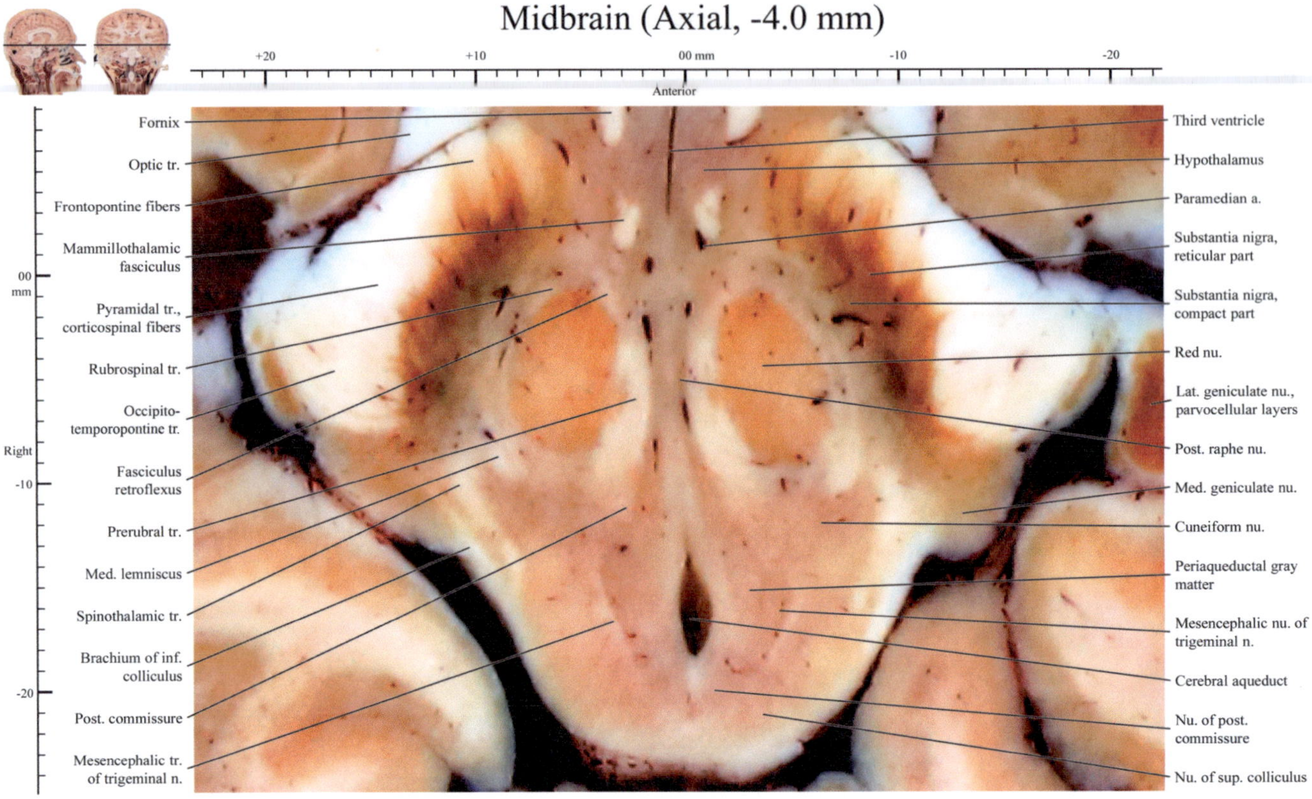

Midbrain (Axial, -4.4 mm)

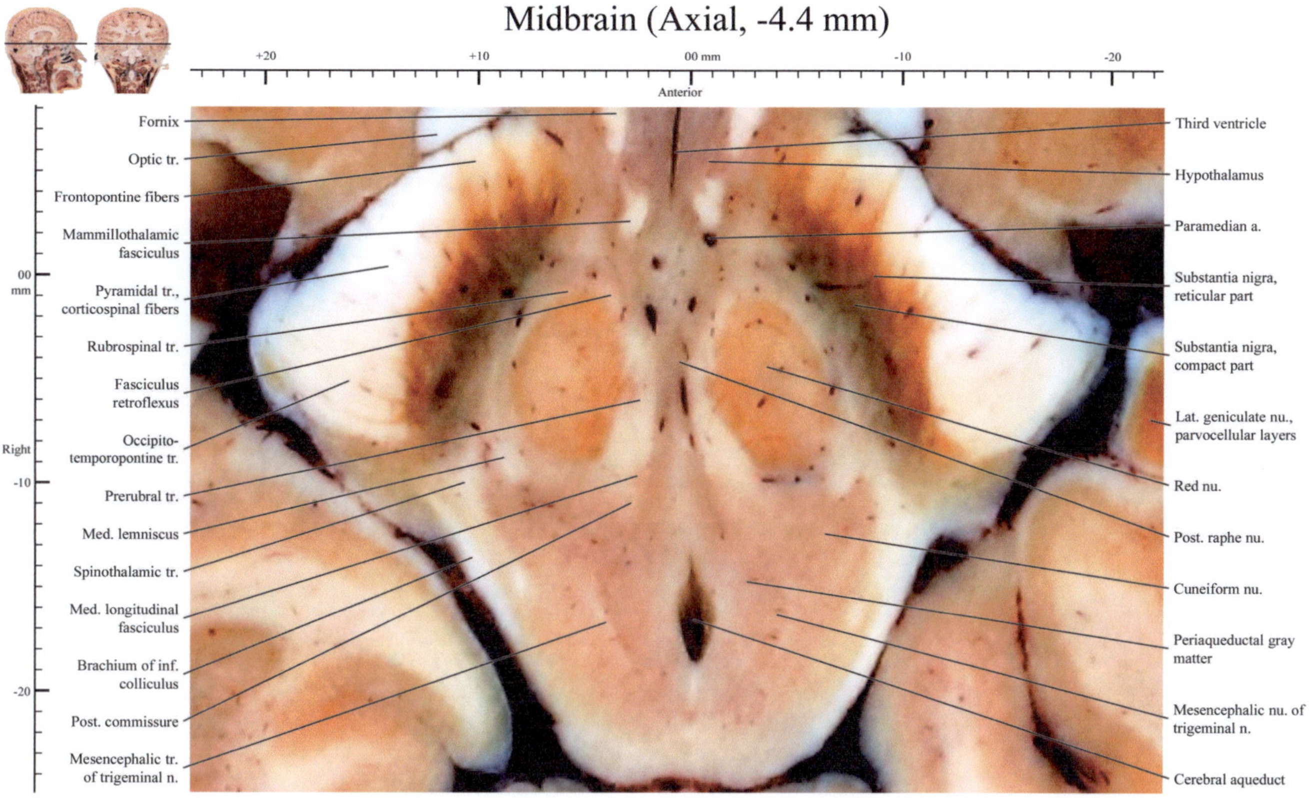

Midbrain (Axial, -5.0 mm)

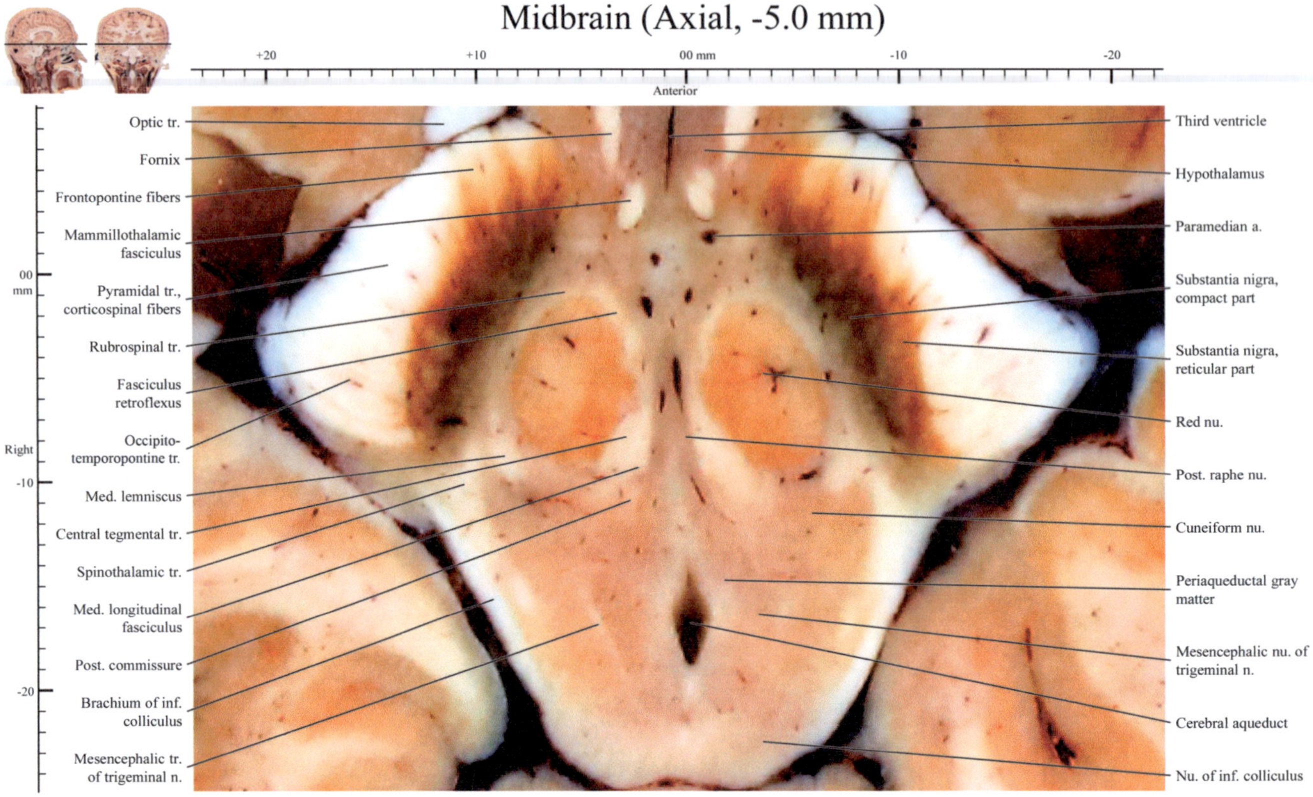

Midbrain (Axial, -5.4 mm)

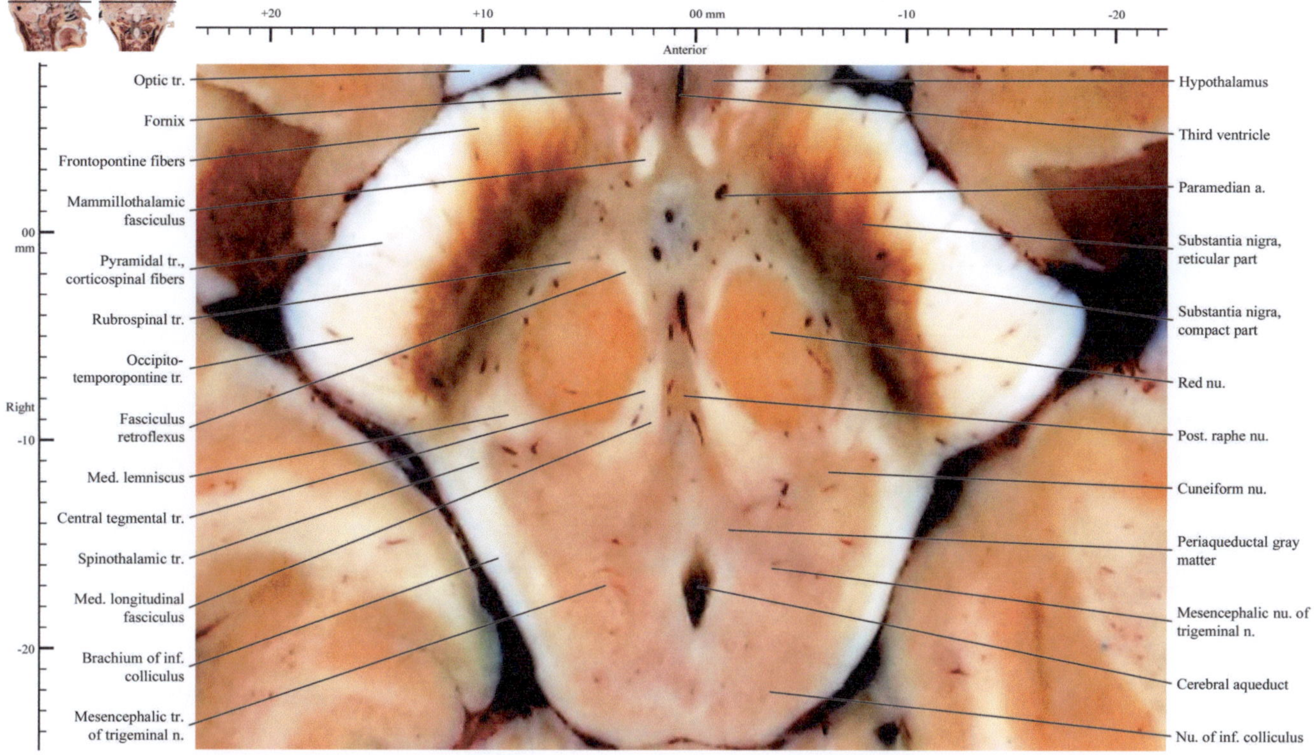

Midbrain (Axial, -6.0 mm)

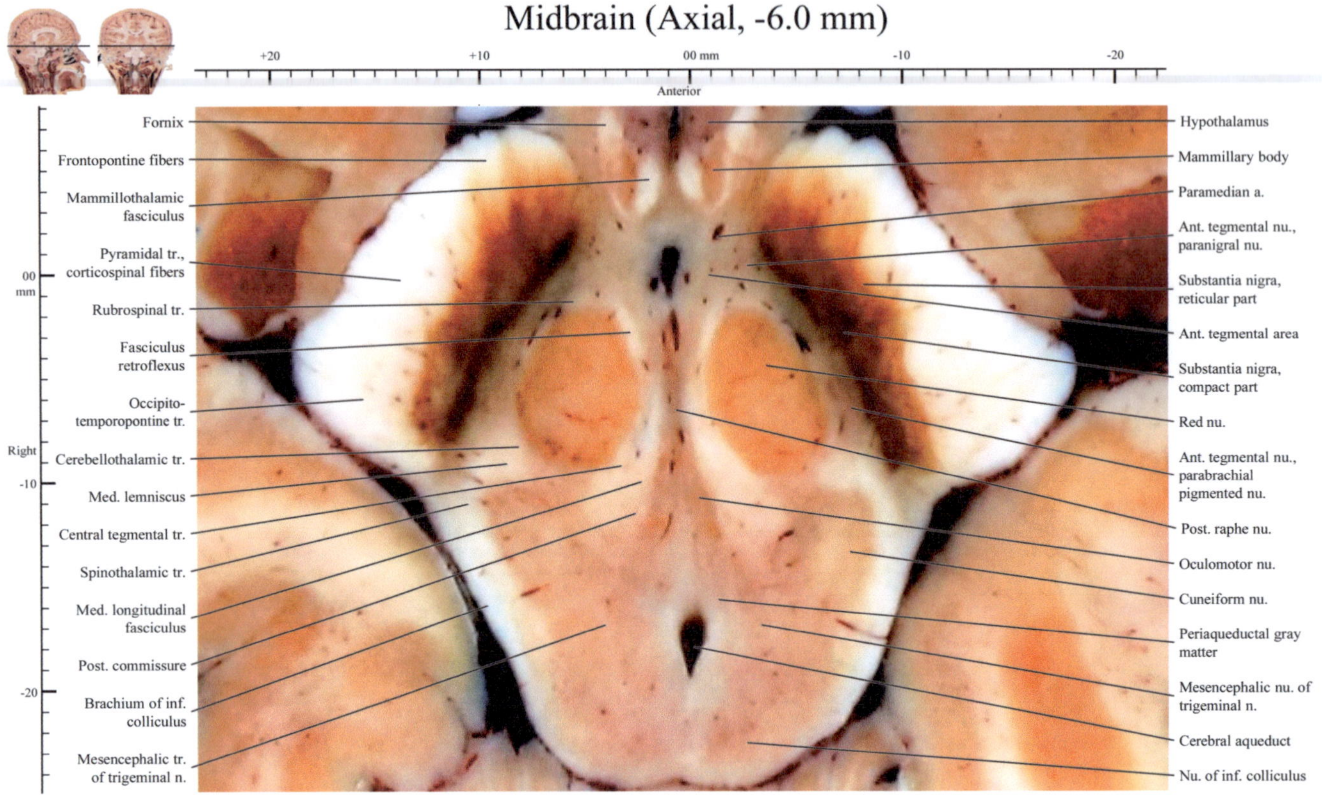

Midbrain (Axial, -6.4 mm)

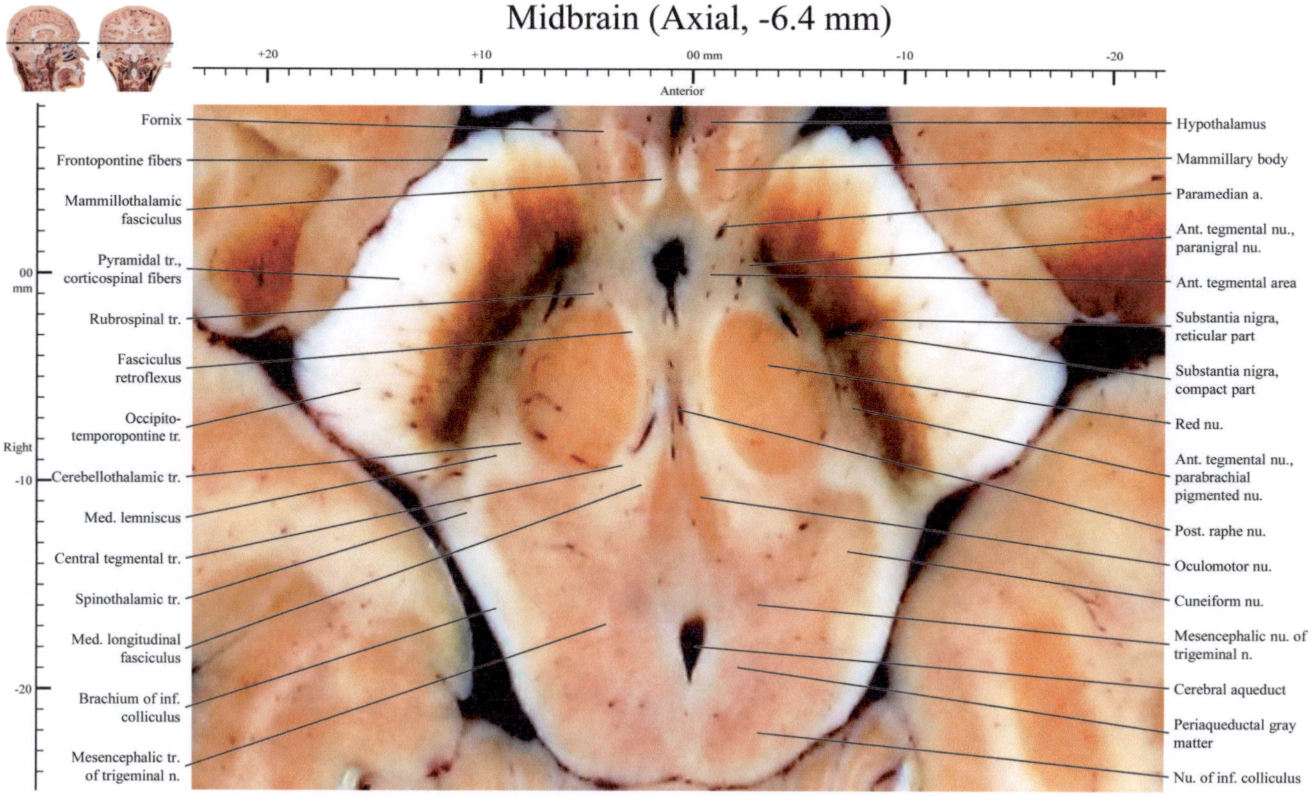

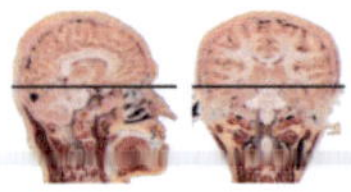

Midbrain (Axial, -7.0 mm)

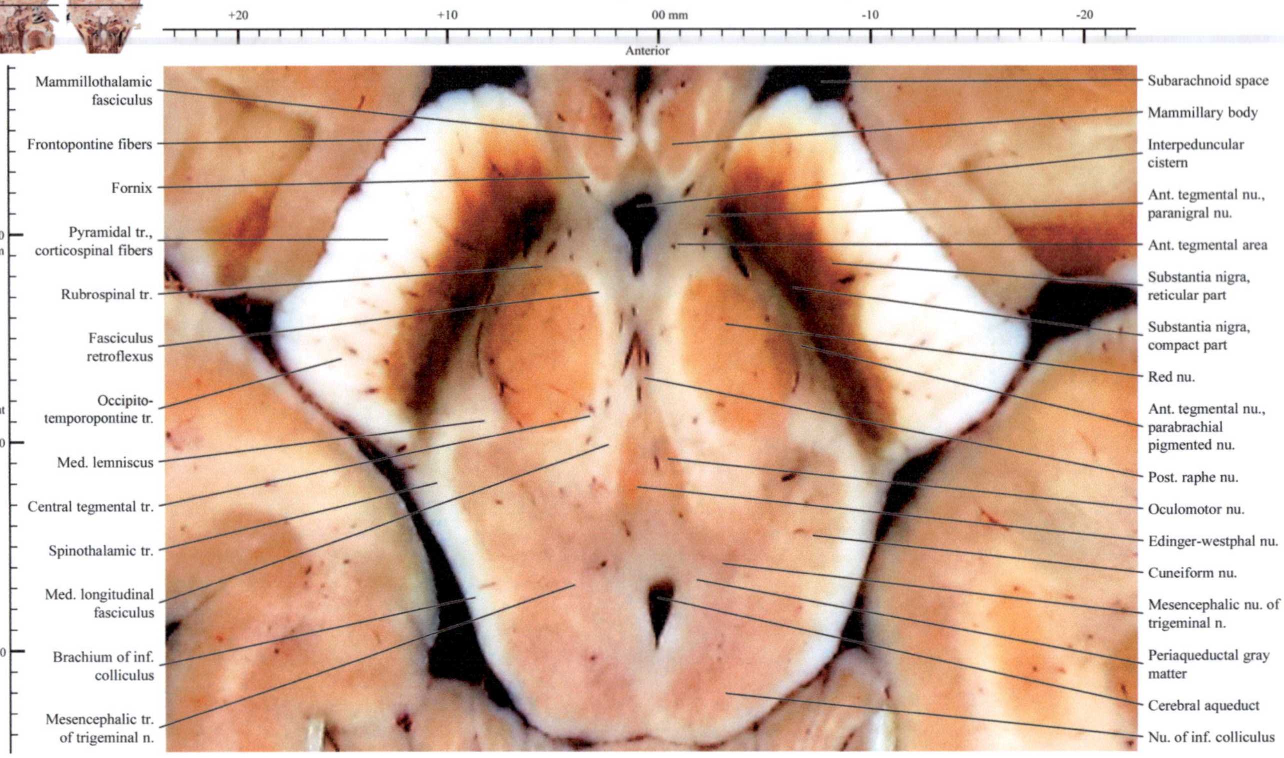

Midbrain (Axial, -7.4 mm)

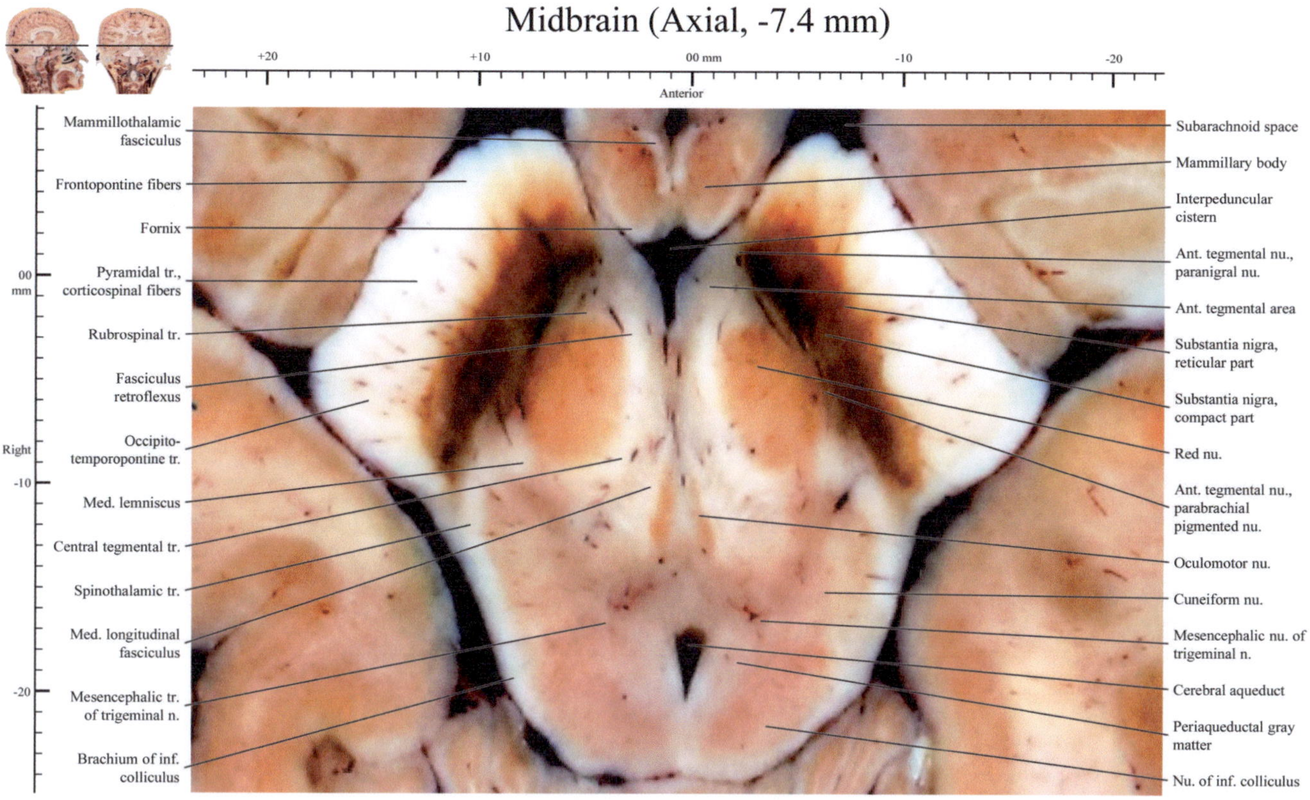

Midbrain (Axial, -8.0 mm)

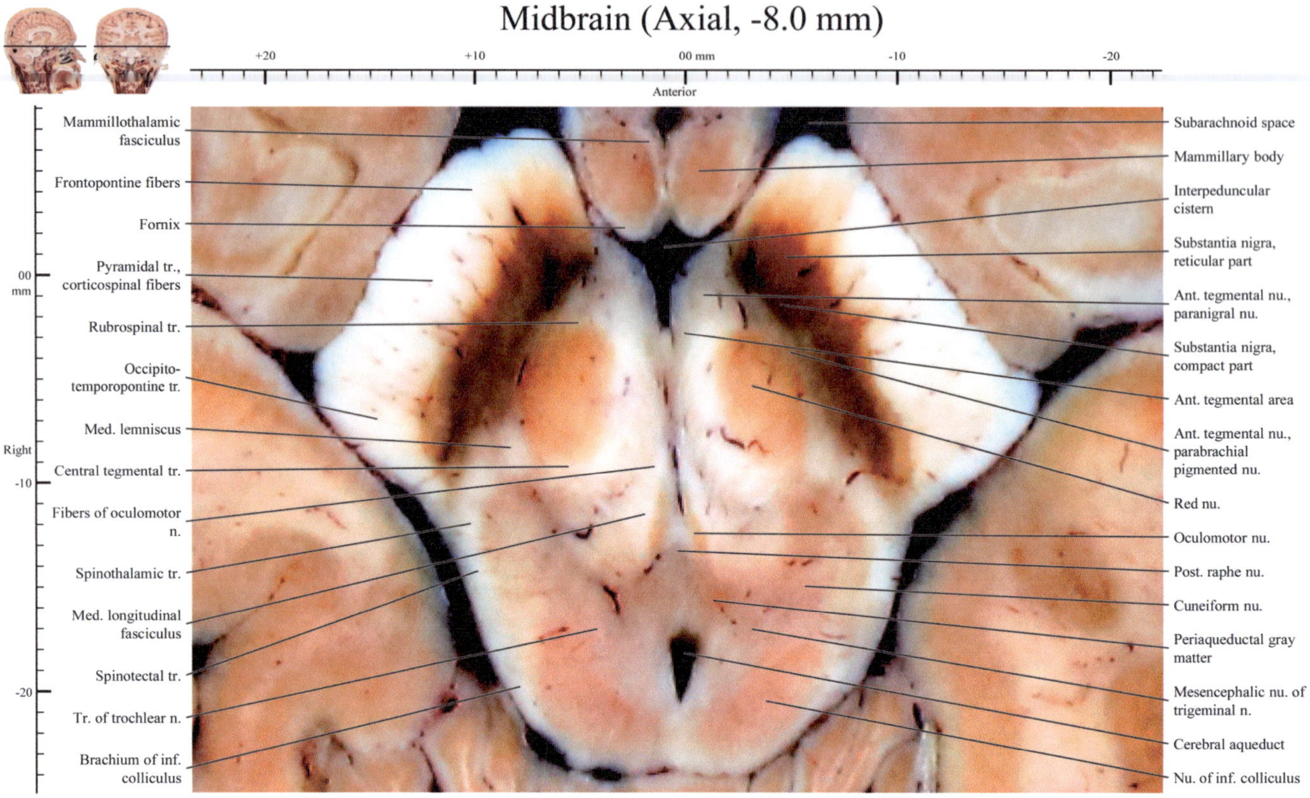

Midbrain (Axial, -8.4 mm)

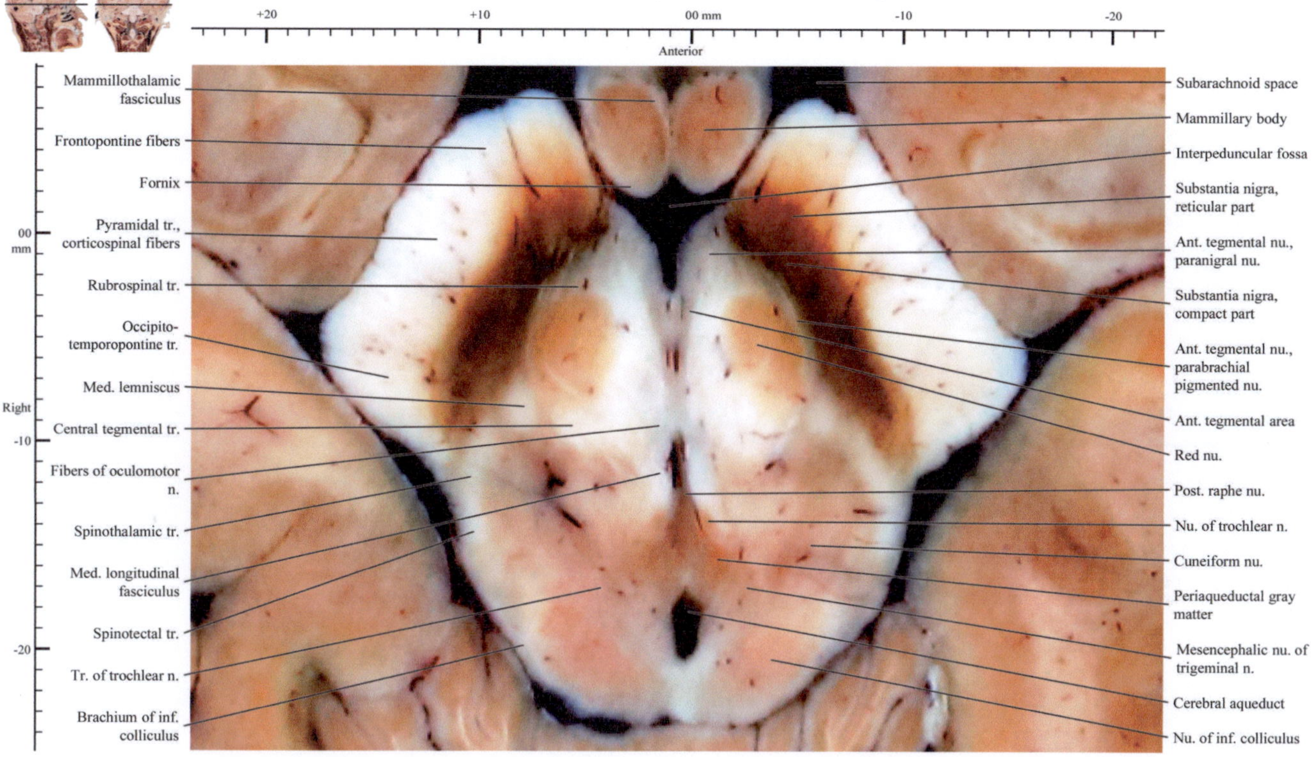

Midbrain (Axial, -9.0 mm)

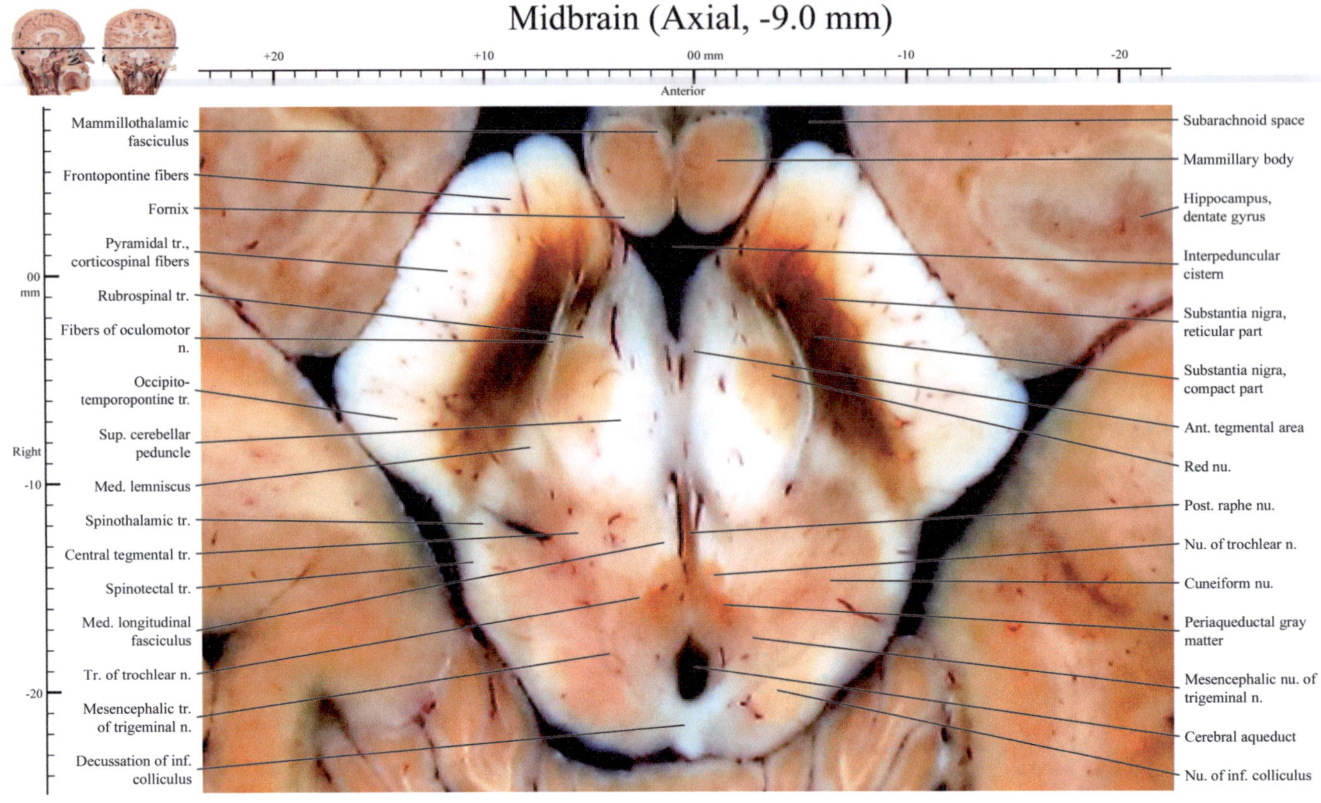

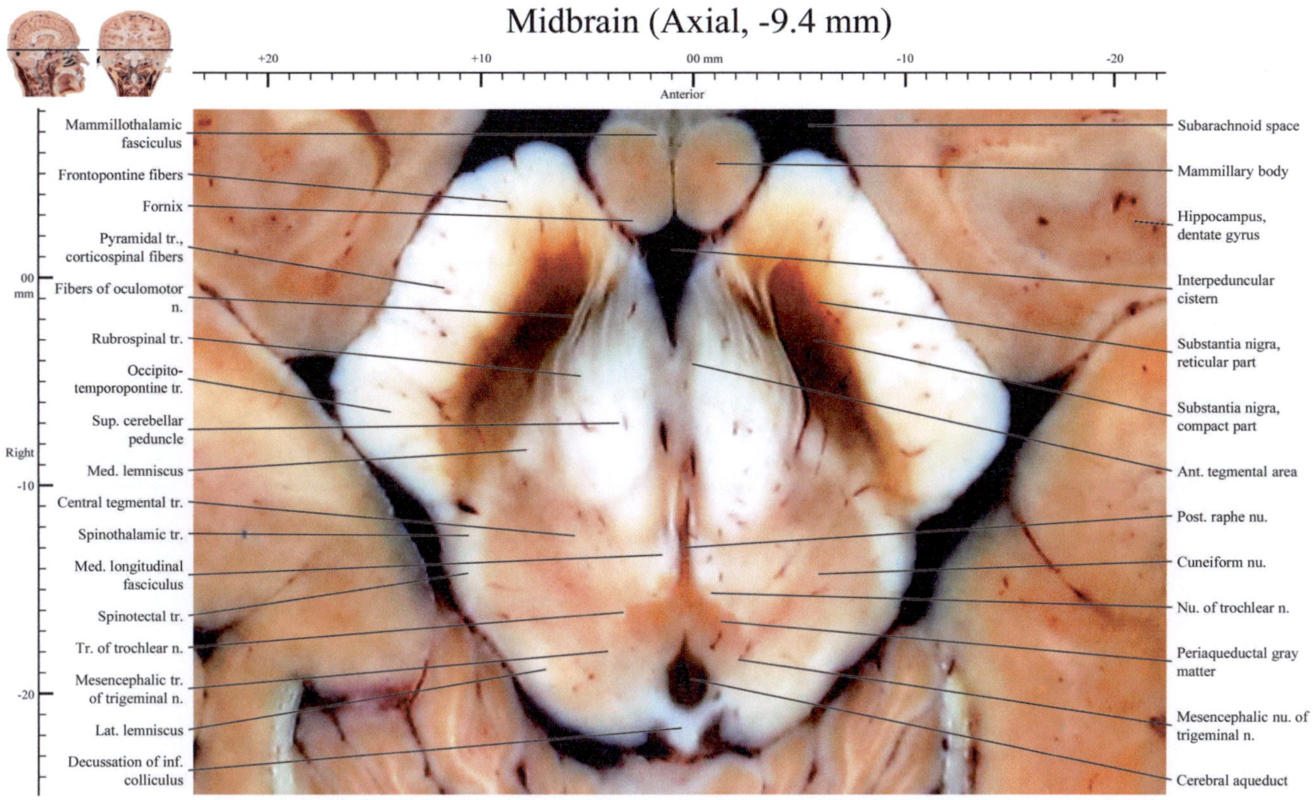
Midbrain (Axial, -9.4 mm)
+20
+10
00 mm
-10
-20
Anterior
00 mm
Right
-10
-20
Mammillothalamic fasciculus
Frontopontine fibers
Fornix
Pyramidal tr., corticospinal fibers
Fibers of oculomotor n.
Rubrospinal tr.
Occipito-temporopontine tr.
Sup. cerebellar peduncle
Med. lemniscus
Central tegmental tr.
Spinothalamic tr.
Med. longitudinal fasciculus
Spinotectal tr.
Tr. of trochlear n.
Mesencephalic tr. of trigeminal n.
Lat. lemniscus
Decussation of inf. colliculus
Subarachnoid space
Mammillary body
Hippocampus, dentate gyrus
Interpeduncular cistern
Substantia nigra, reticular part
Substantia nigra, compact part
Ant. tegmental area
Post. raphe nu.
Cuneiform nu.
Nu. of trochlear n.
Periaqueductal gray matter
Mesencephalic nu. of trigeminal n.
Cerebral aqueduct

Midbrain (Axial, -10.0 mm)

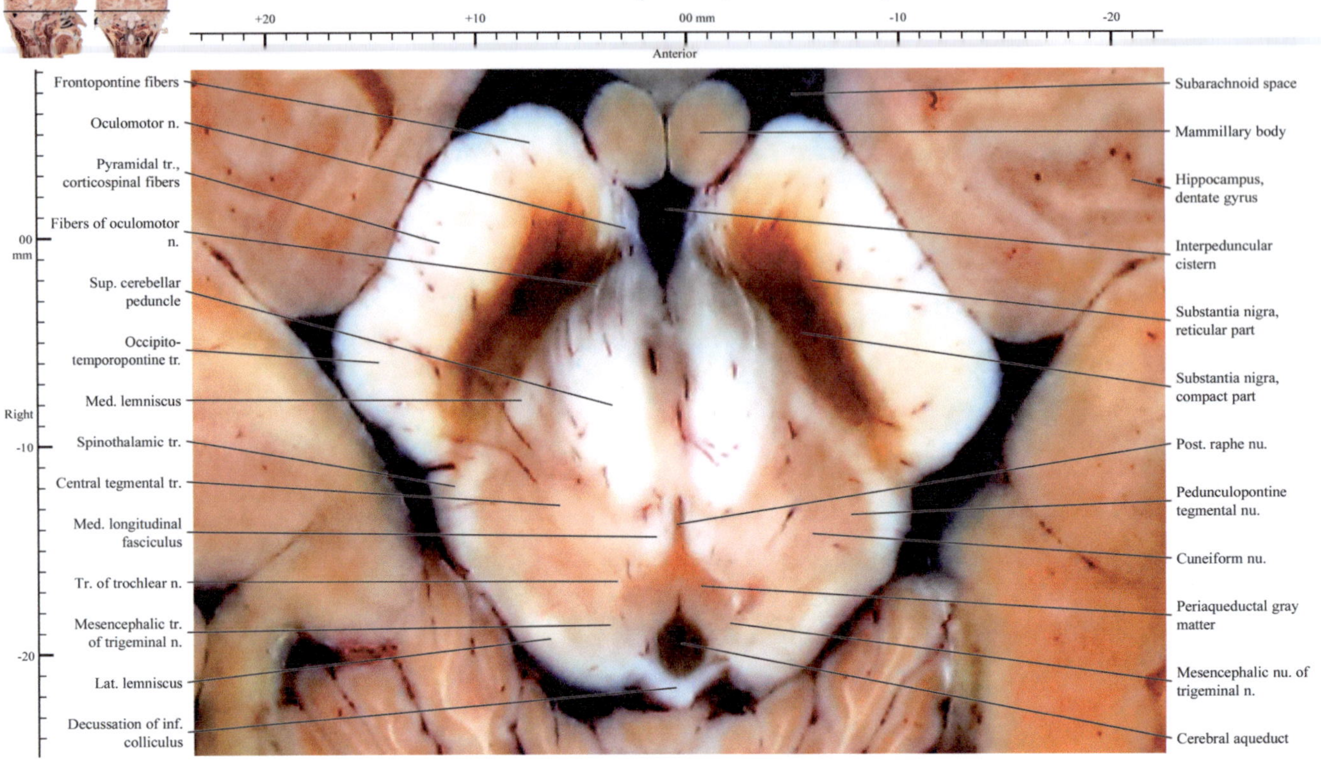

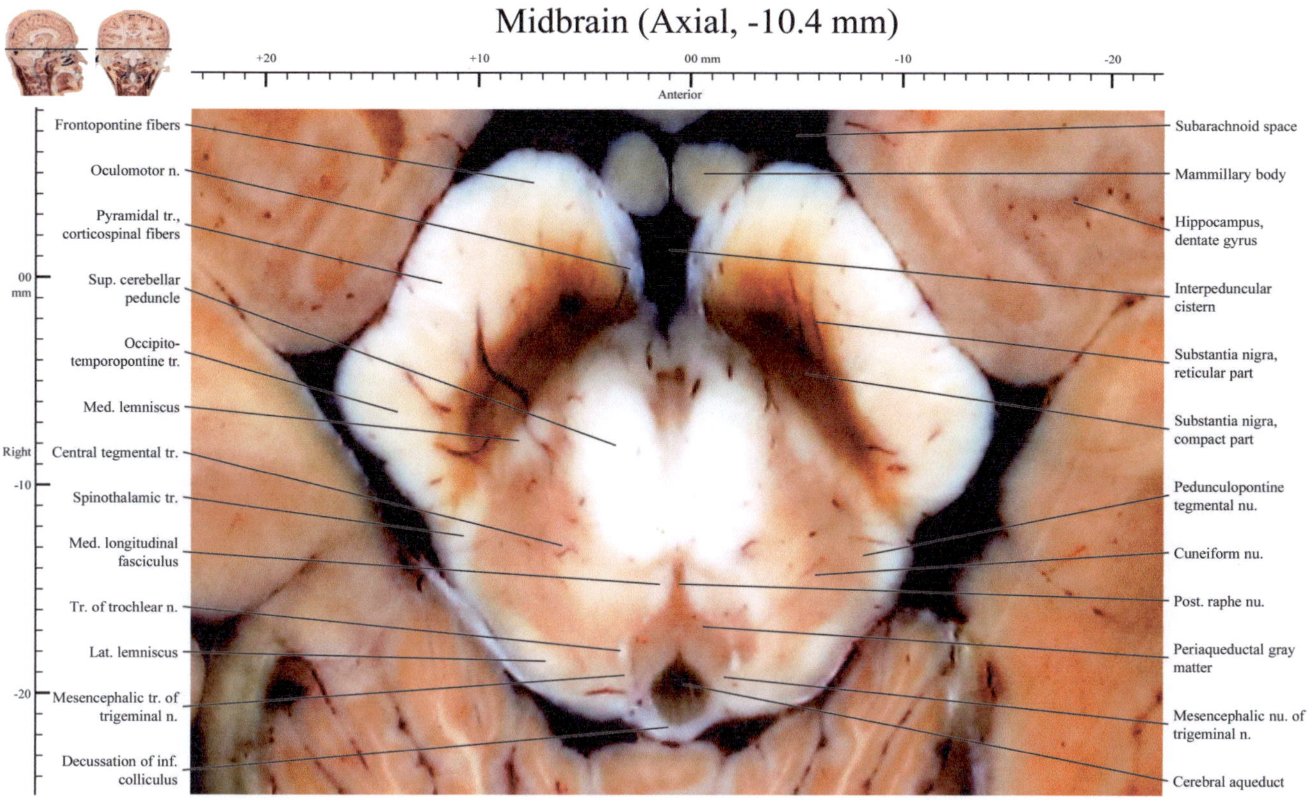

Midbrain (Axial, -10.4 mm)
+20
+10
00 mm
-10
-20
Anterior
00 mm
Right
-10
-20
Frontopontine fibers
Oculomotor n.
Pyramidal tr., corticospinal fibers
Sup. cerebellar peduncle
Occipito-temporopontine tr.
Med. lemniscus
Central tegmental tr.
Spinothalamic tr.
Med. longitudinal fasciculus
Tr. of trochlear n.
Lat. lemniscus
Mesencephalic tr. of trigeminal n.
Decussation of inf. colliculus
Subarachnoid space
Mammillary body
Hippocampus, dentate gyrus
Interpeduncular cistern
Substantia nigra, reticular part
Substantia nigra, compact part
Pedunculopontine tegmental nu.
Cuneiform nu.
Post. raphe nu.
Periaqueductal gray matter
Mesencephalic nu. of trigeminal n.
Cerebral aqueduct

Midbrain (Axial, -11.0 mm)

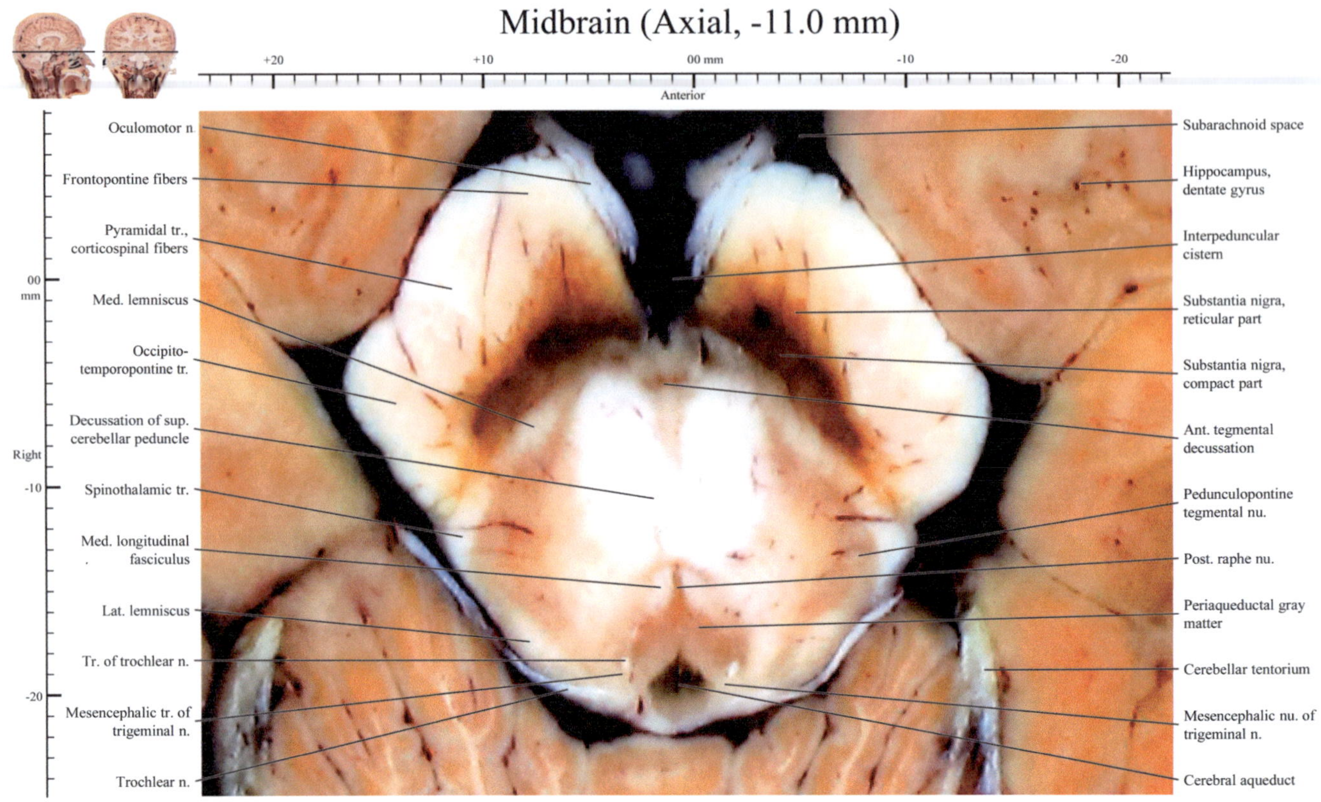

Midbrain (Axial, -11.4 mm)

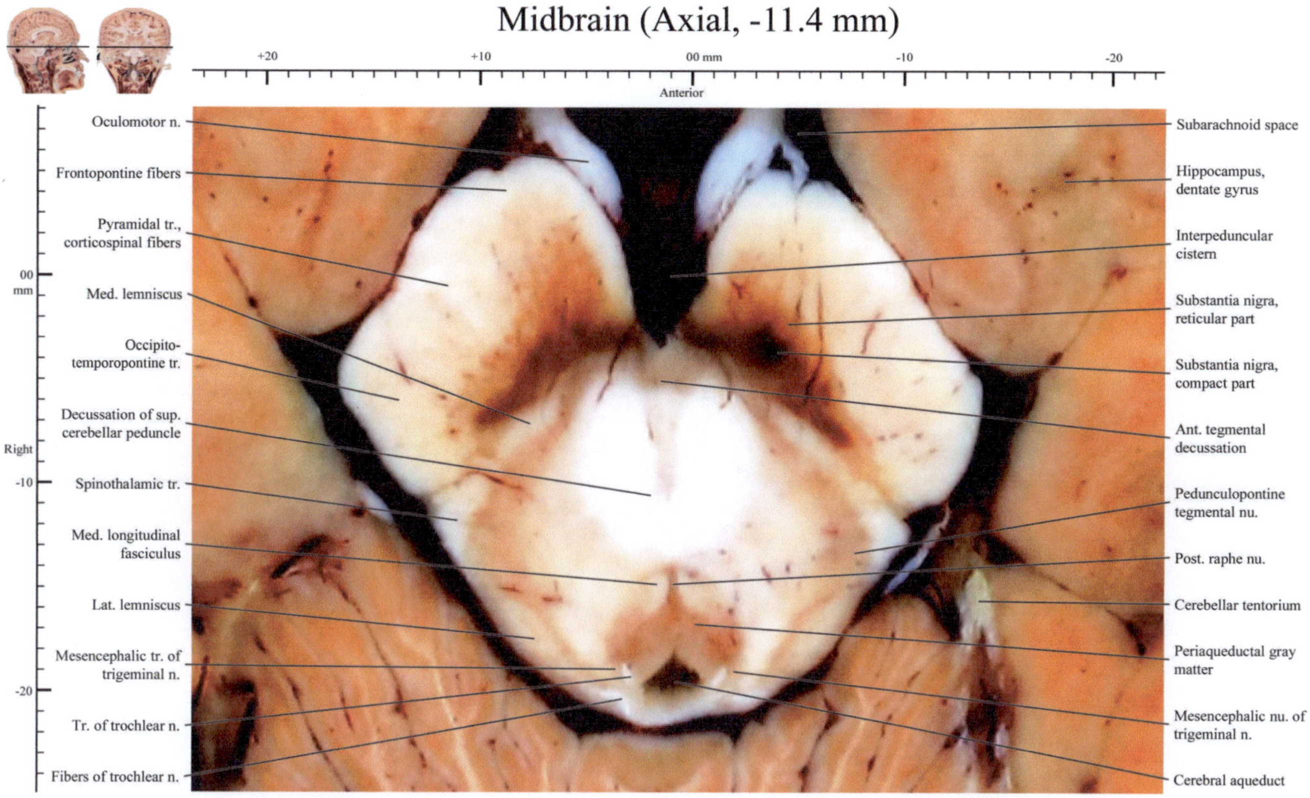

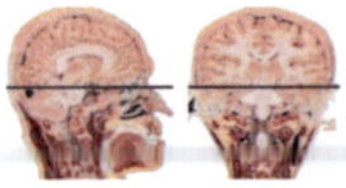

Midbrain (Axial, -12.0 mm)

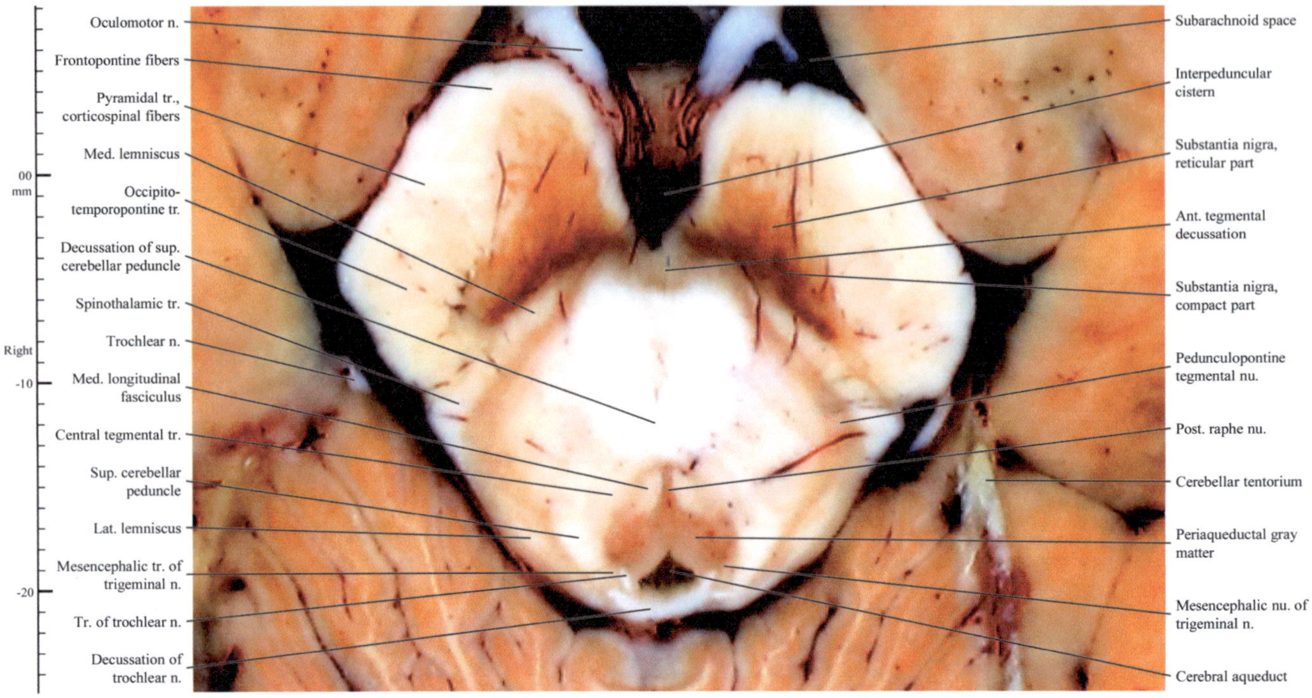

Midbrain (Axial, -12.4 mm)

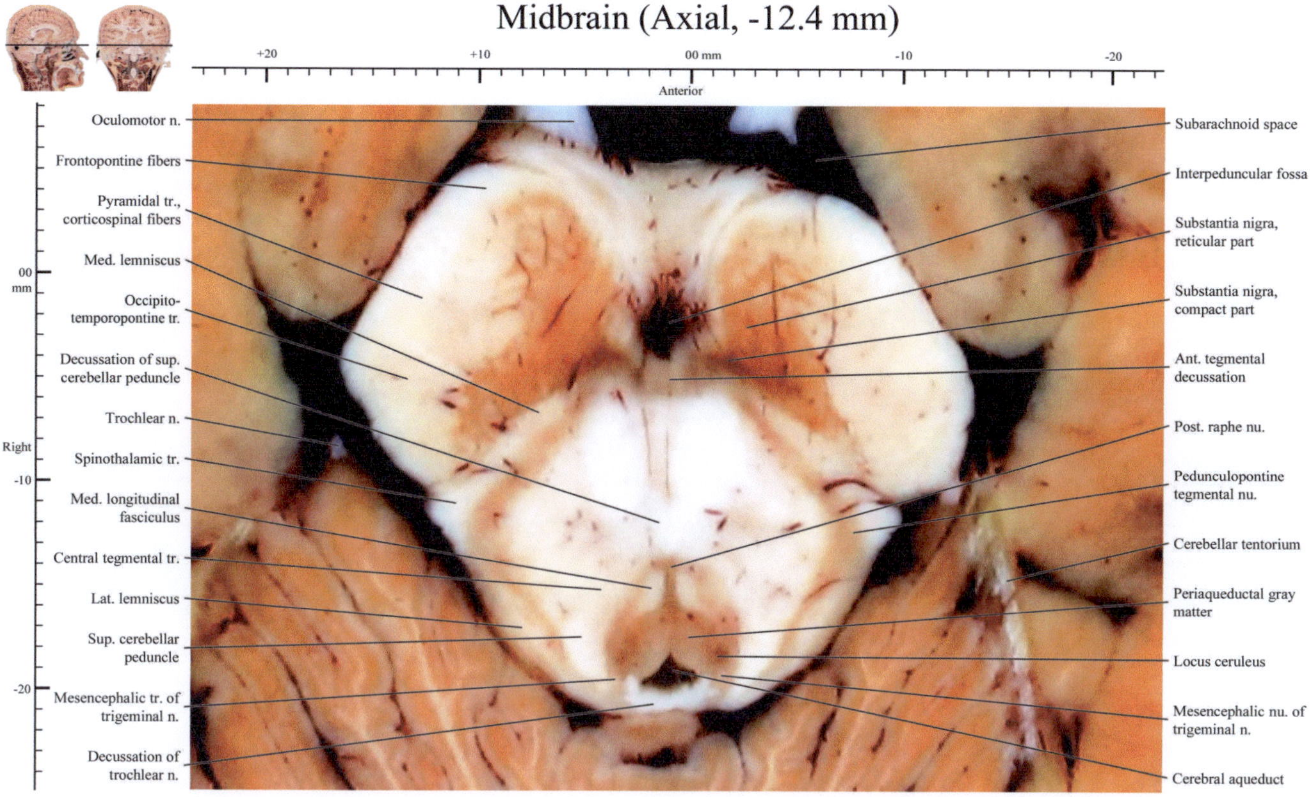

Midbrain and pons (Axial, -13.0 mm)

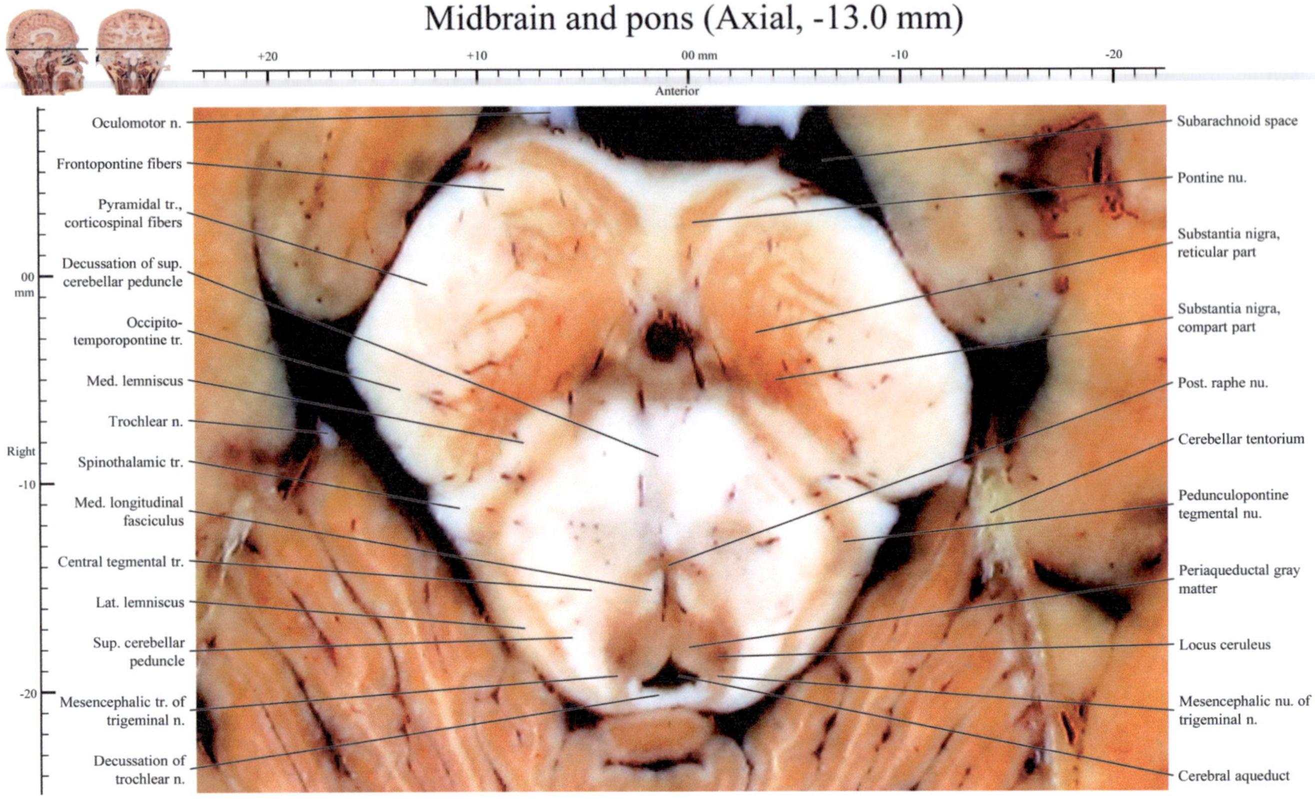

Midbrain and pons (Axial, -13.4 mm)

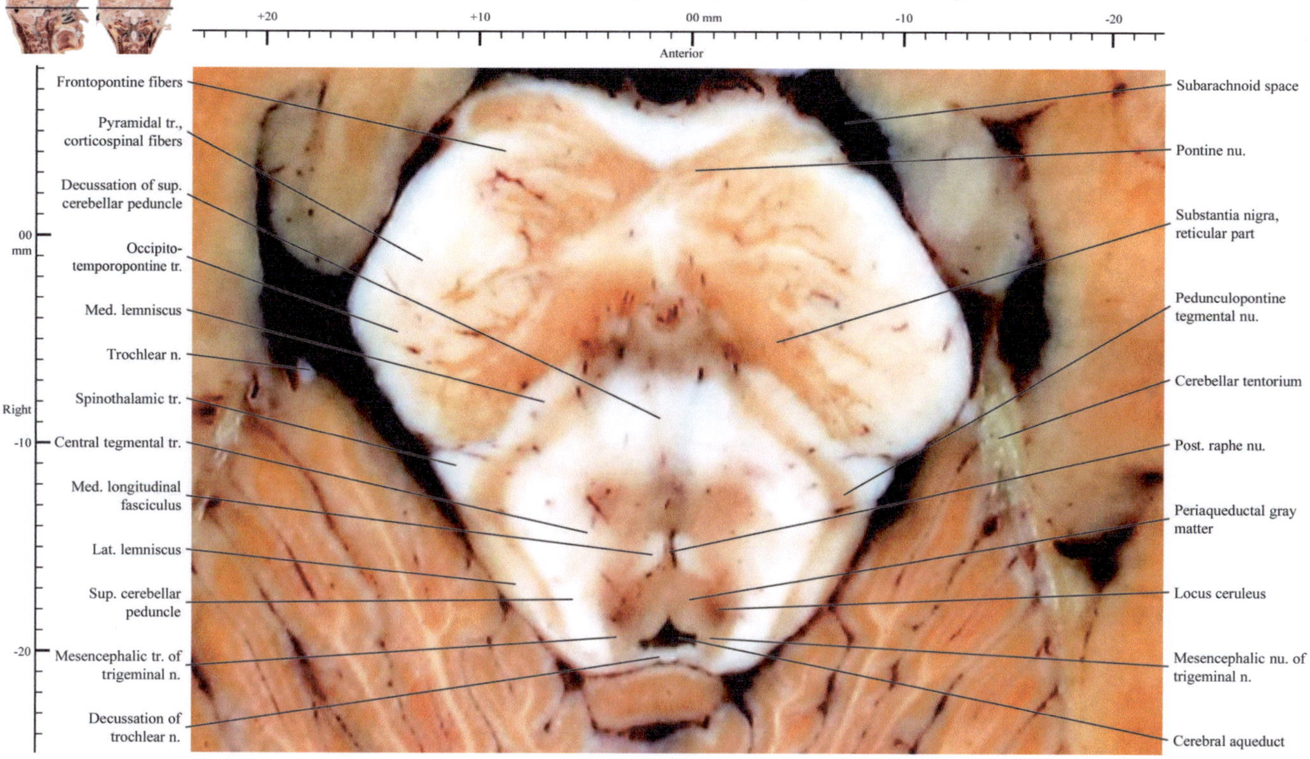

Midbrain and pons (Axial, -14.0 mm)

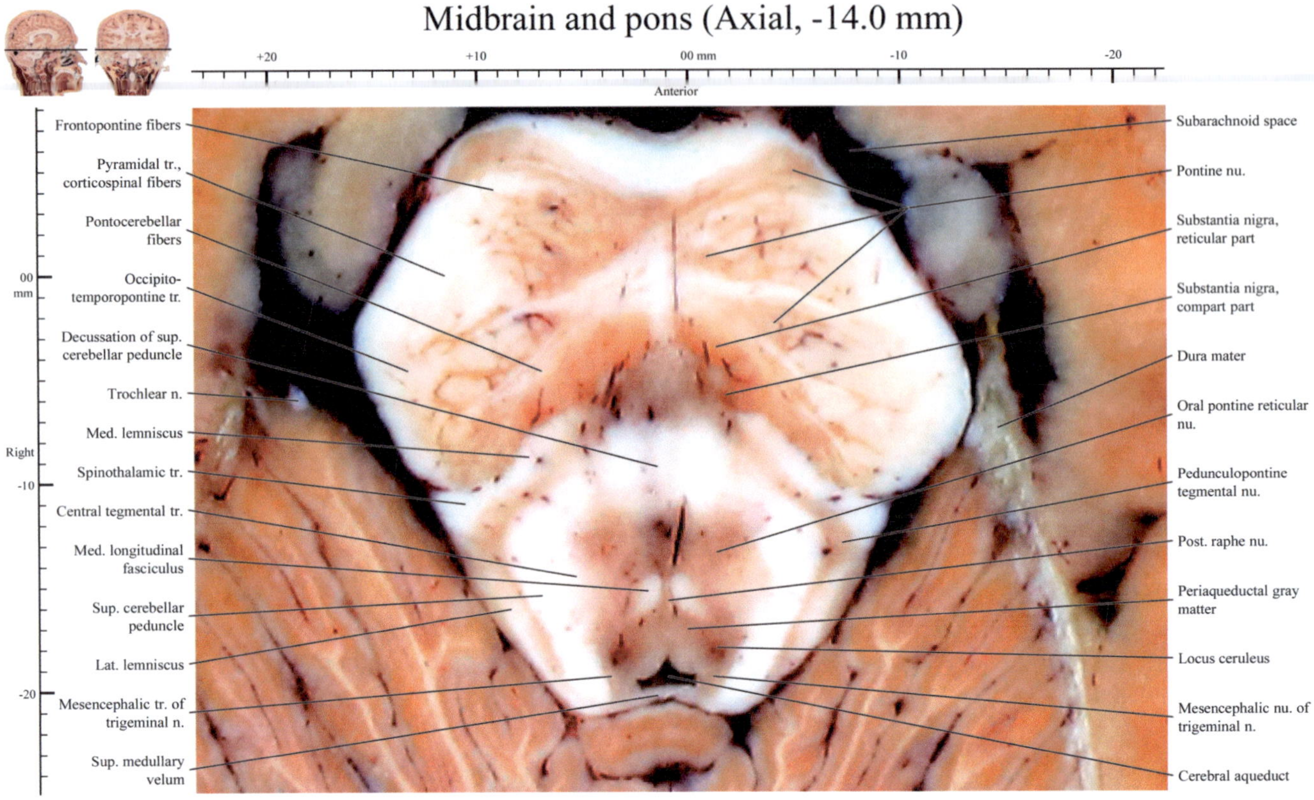

Midbrain and pons (Axial, -14.4 mm)

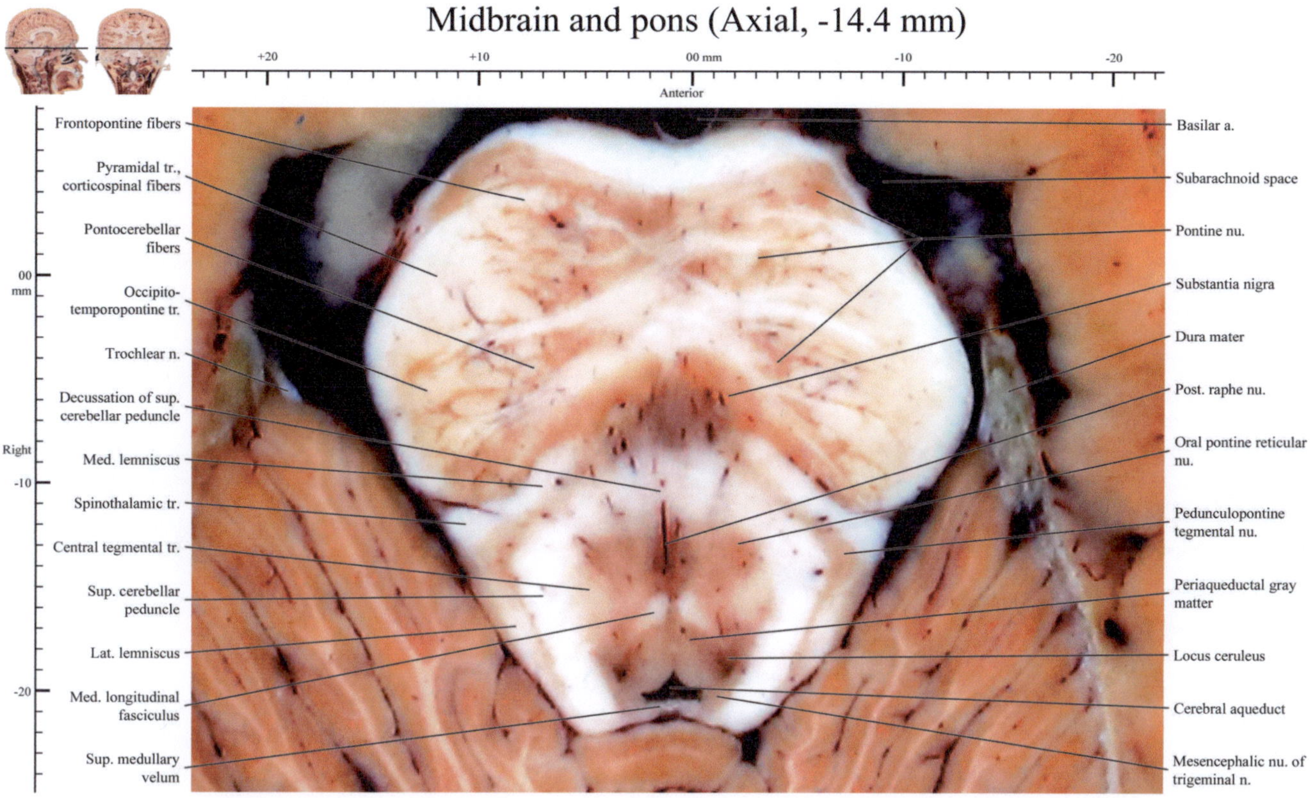

Pons (Axial, -15.0 mm)

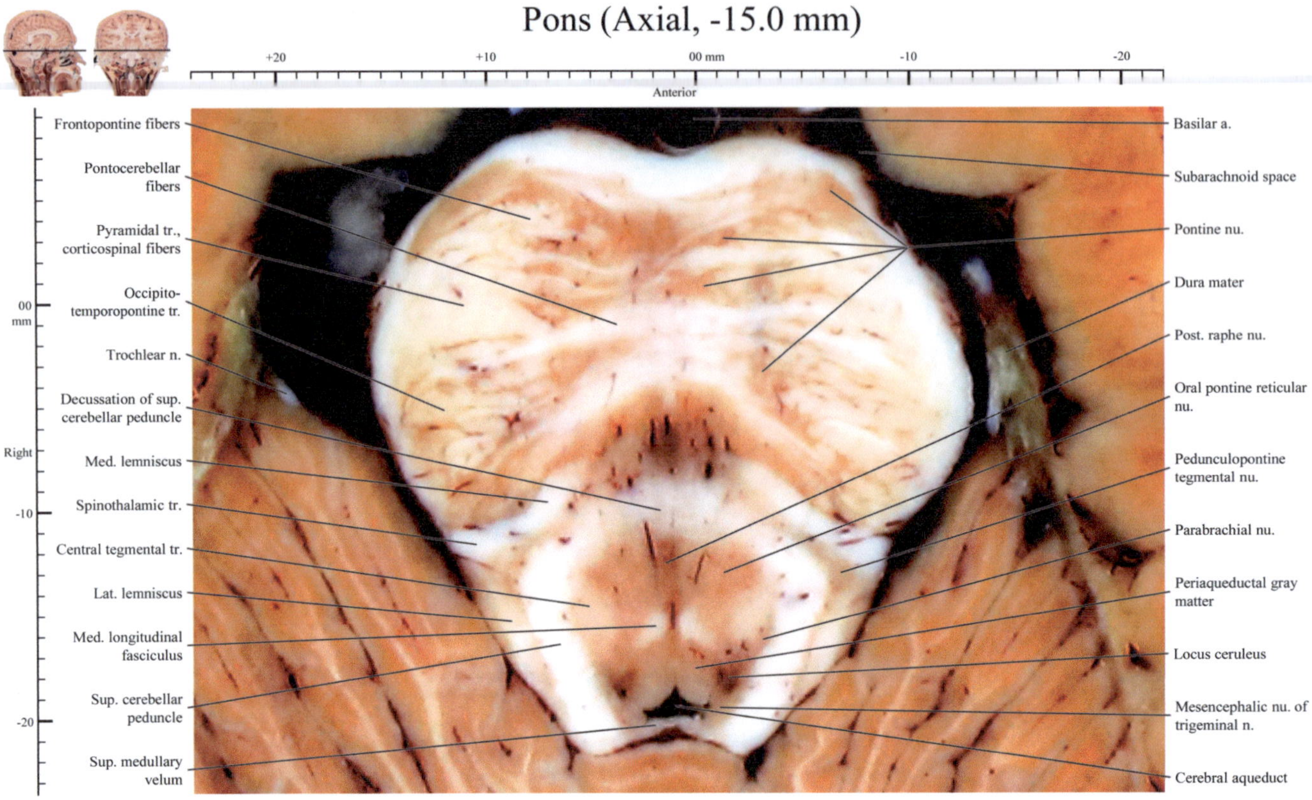

Pons (Axial, -15.4 mm)

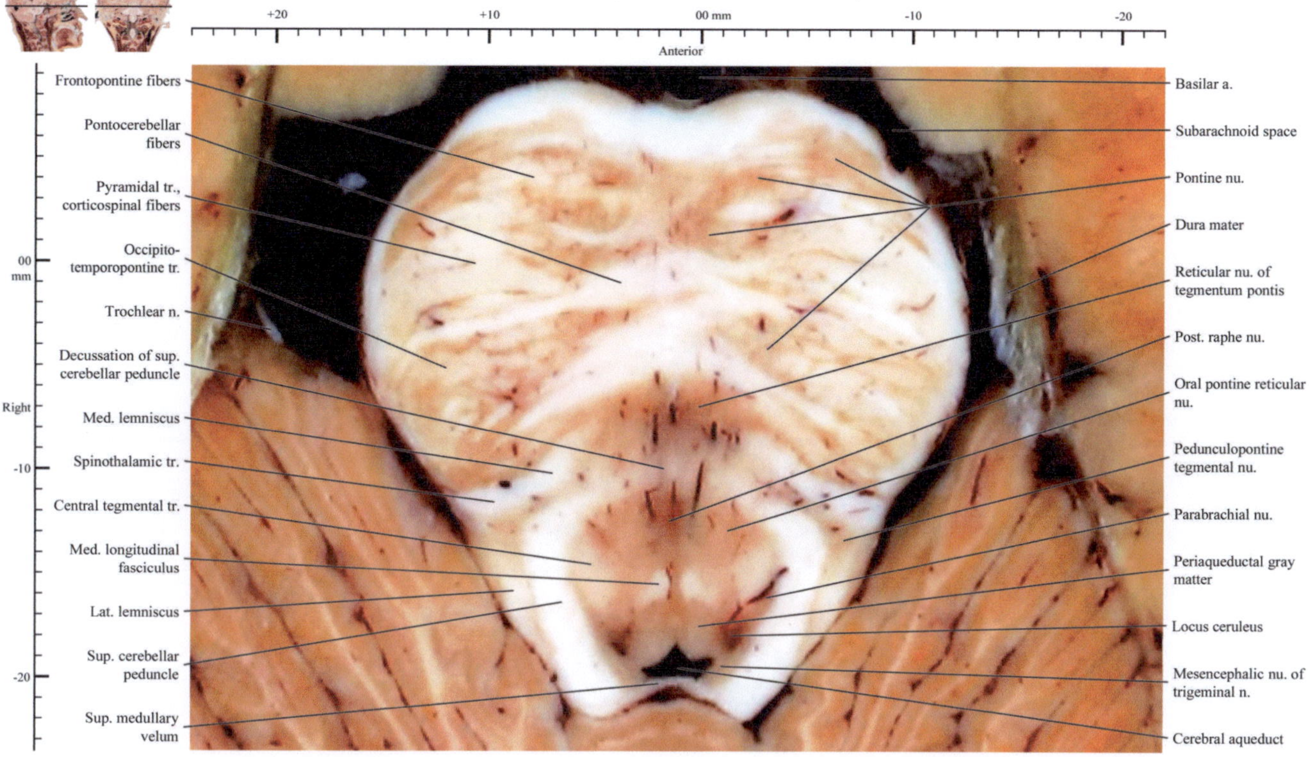

Pons (Axial, -16.0 mm)

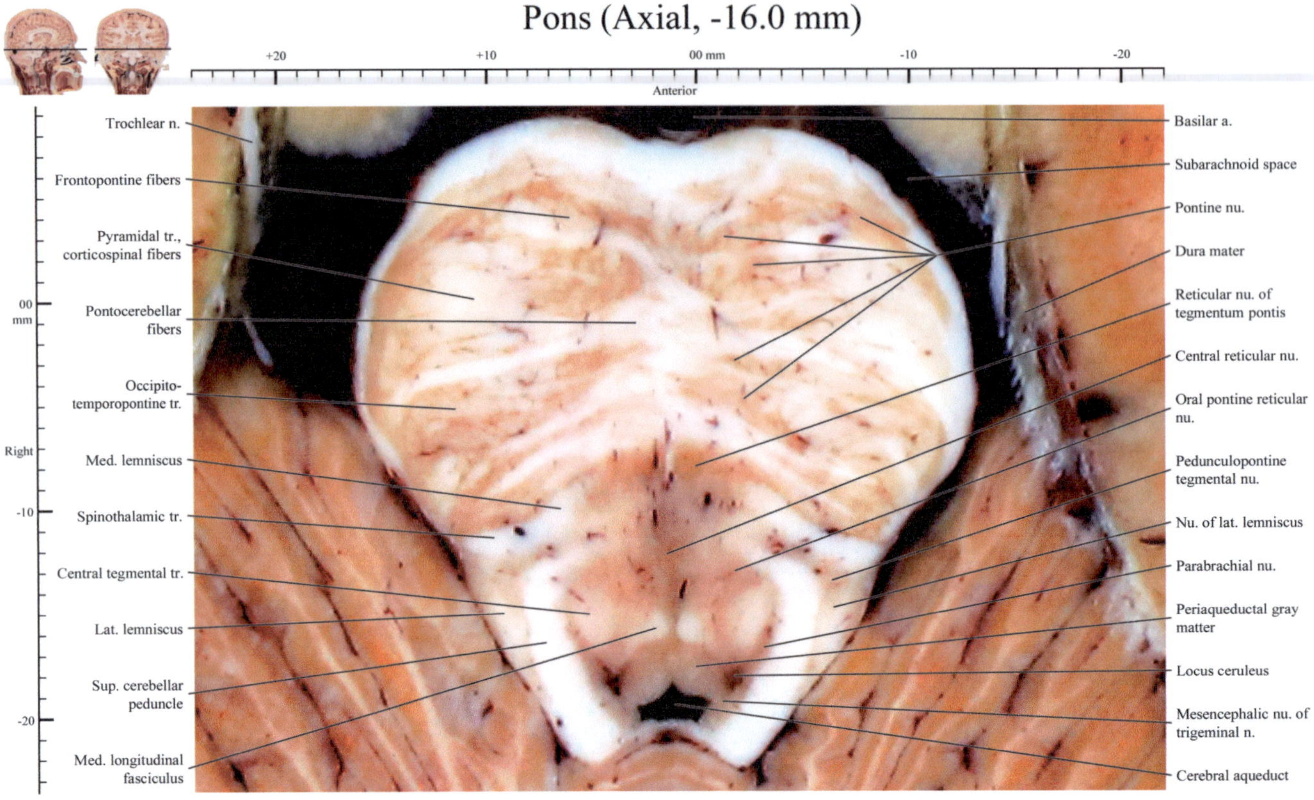

Pons (Axial, -16.4 mm)

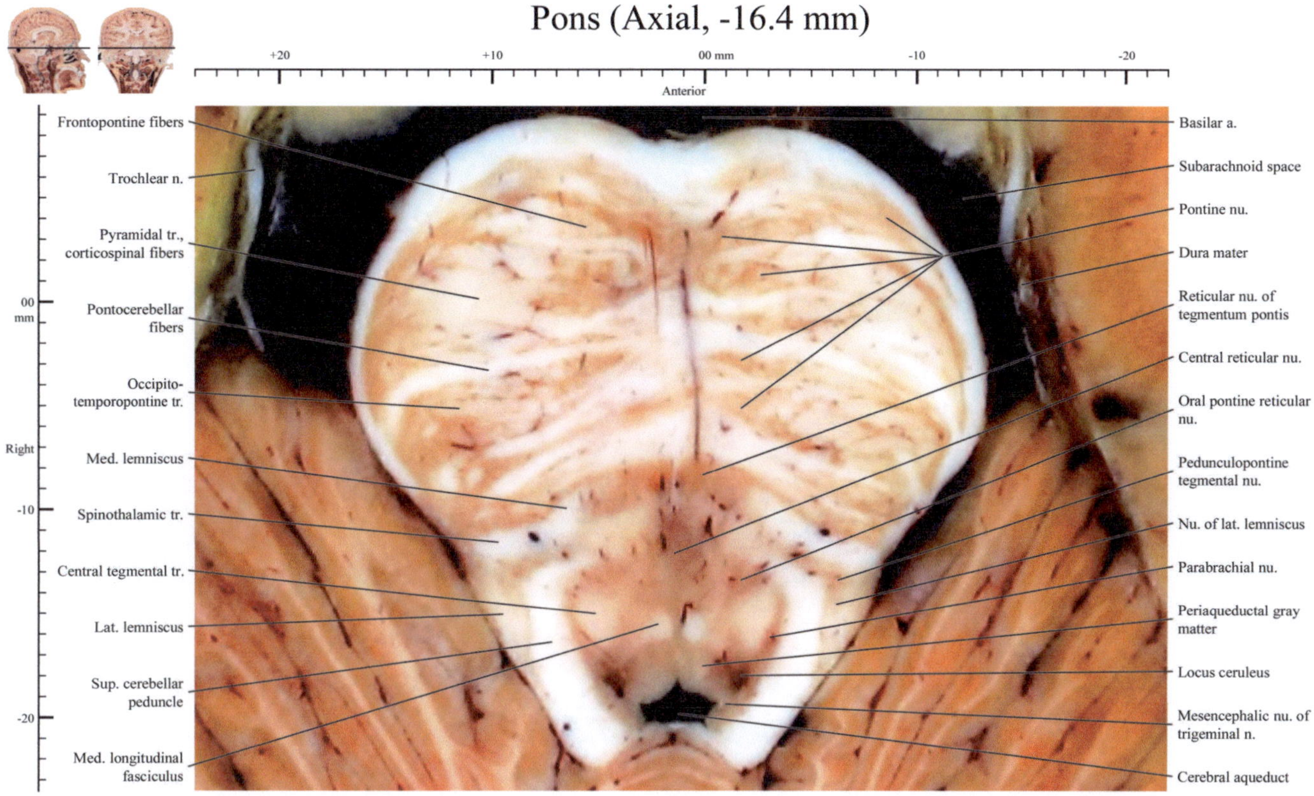

Pons (Axial, -17.0 mm)

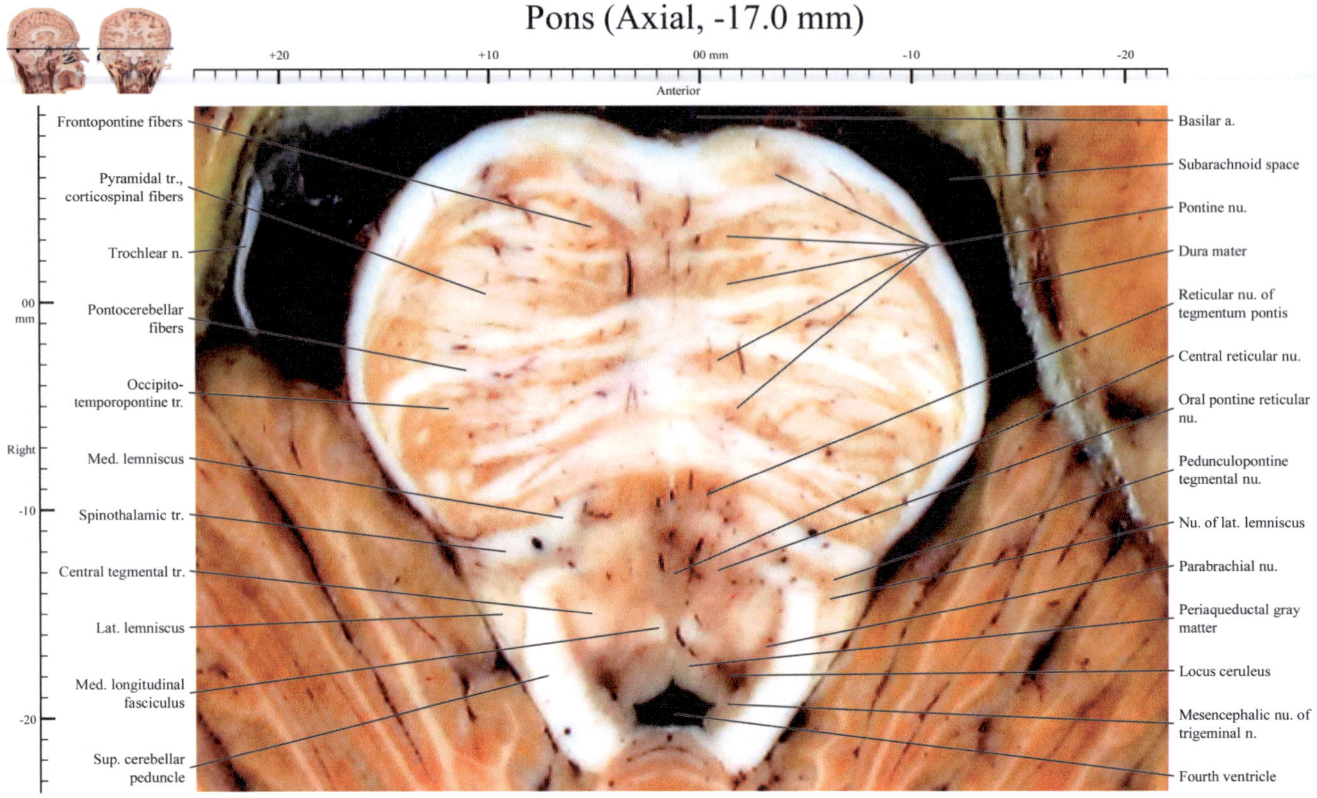

Pons (Axial, -17.4 mm)

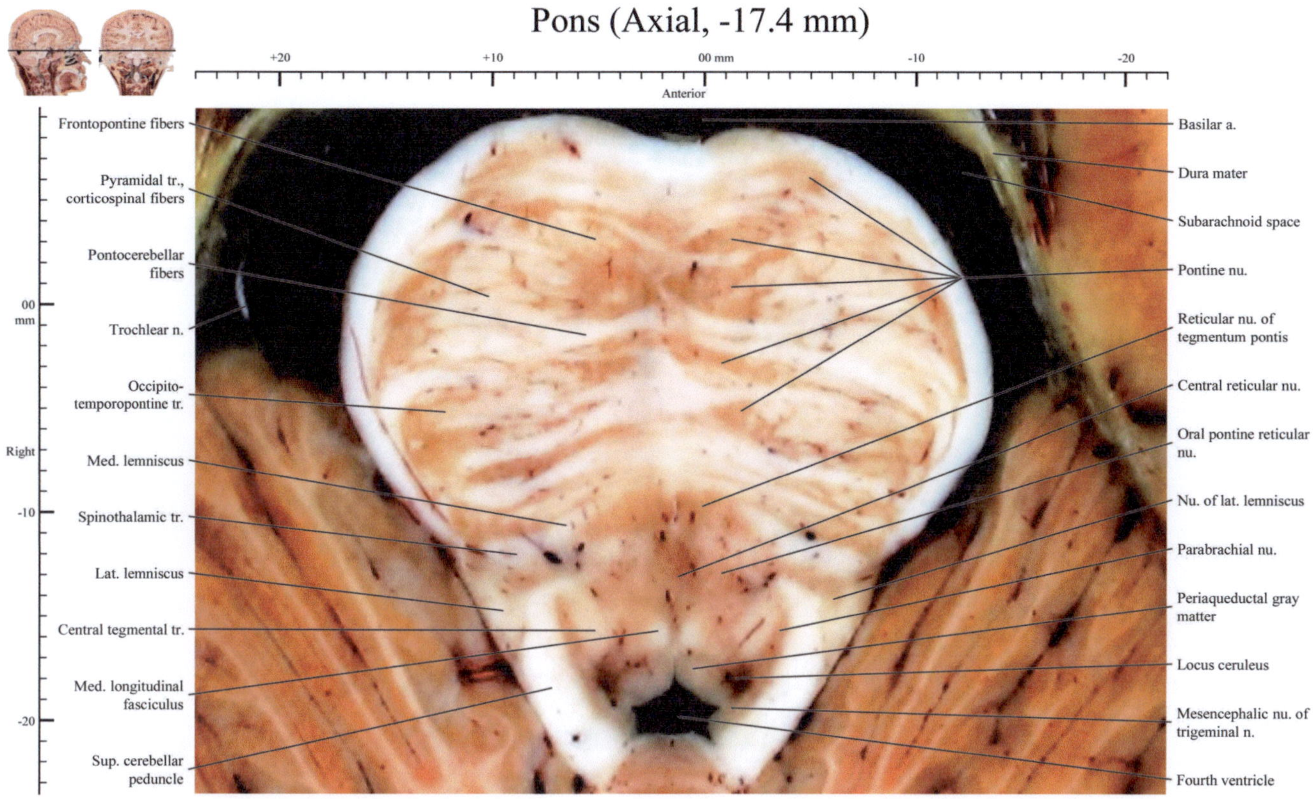

Pons (Axial, -18.0 mm)

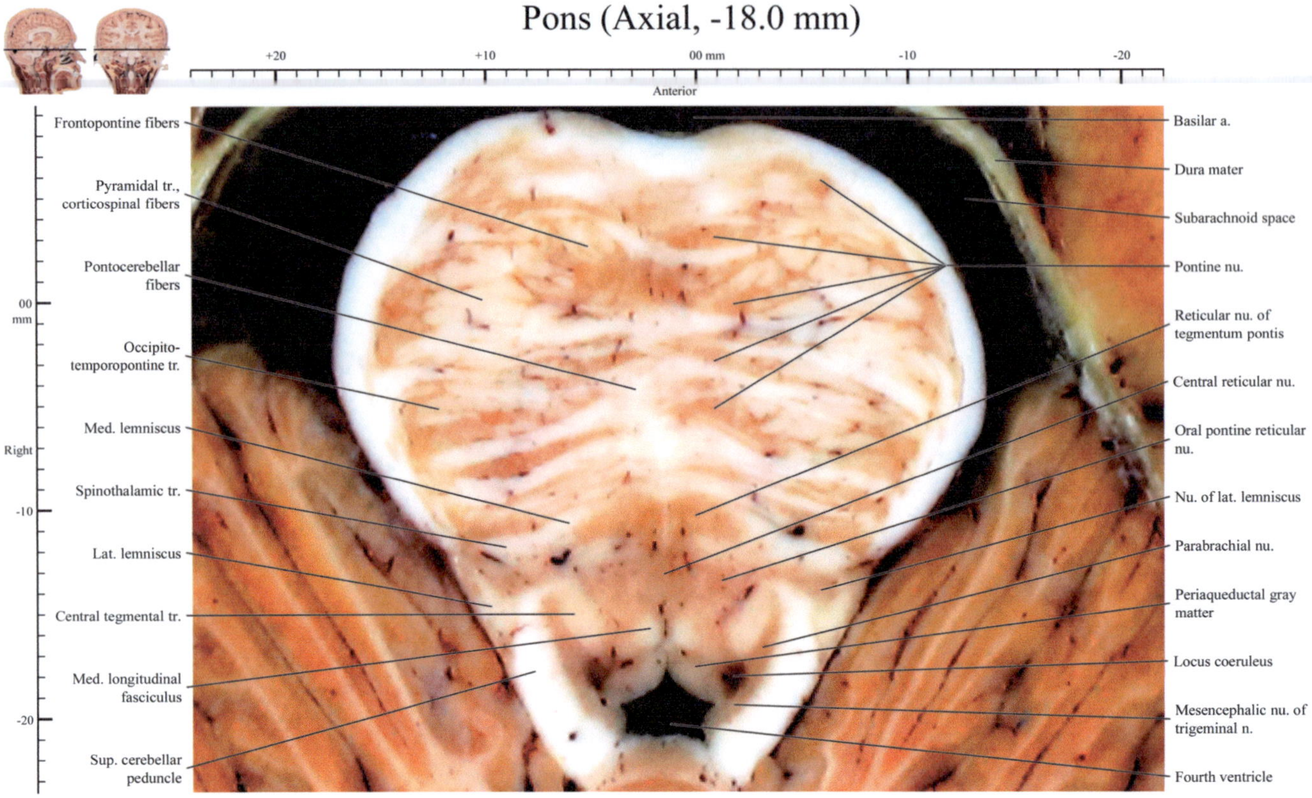

Pons (Axial, -18.4 mm)

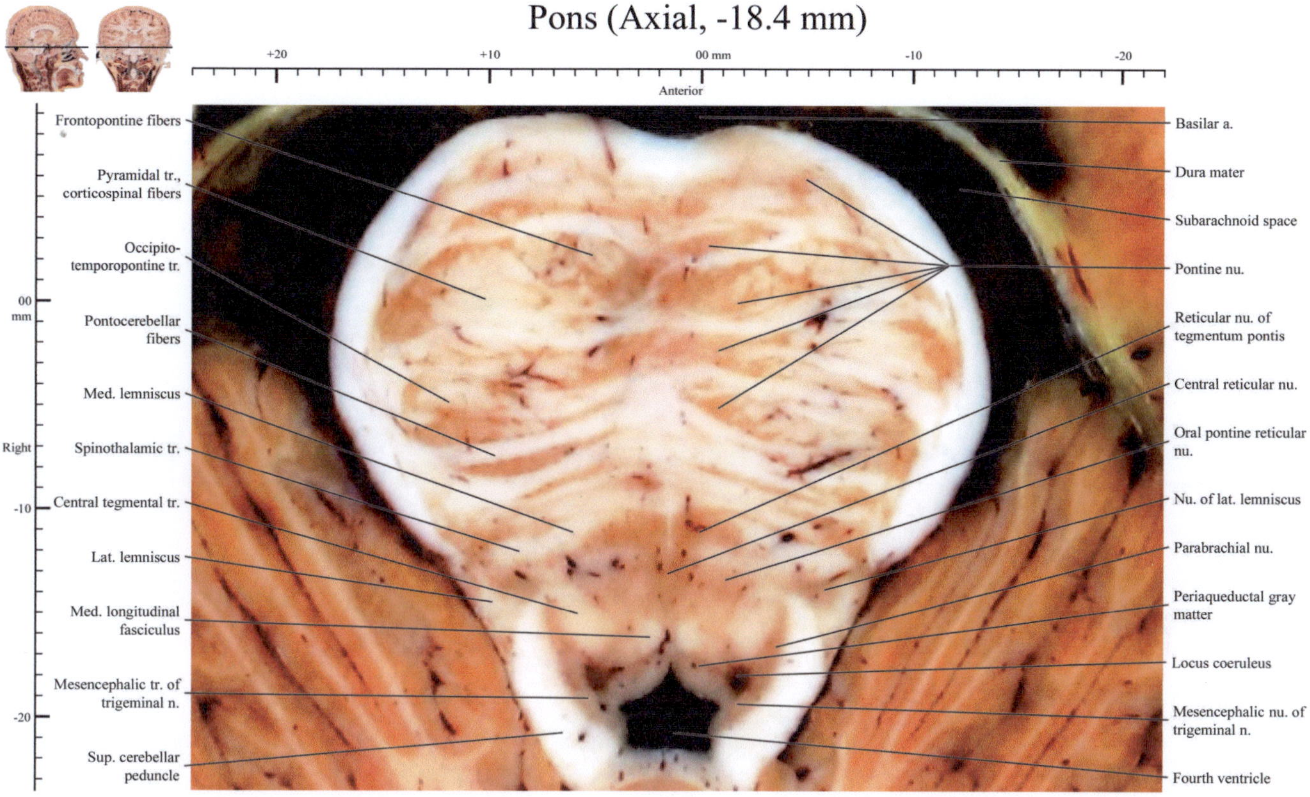

Pons (Axial, -19.0 mm)

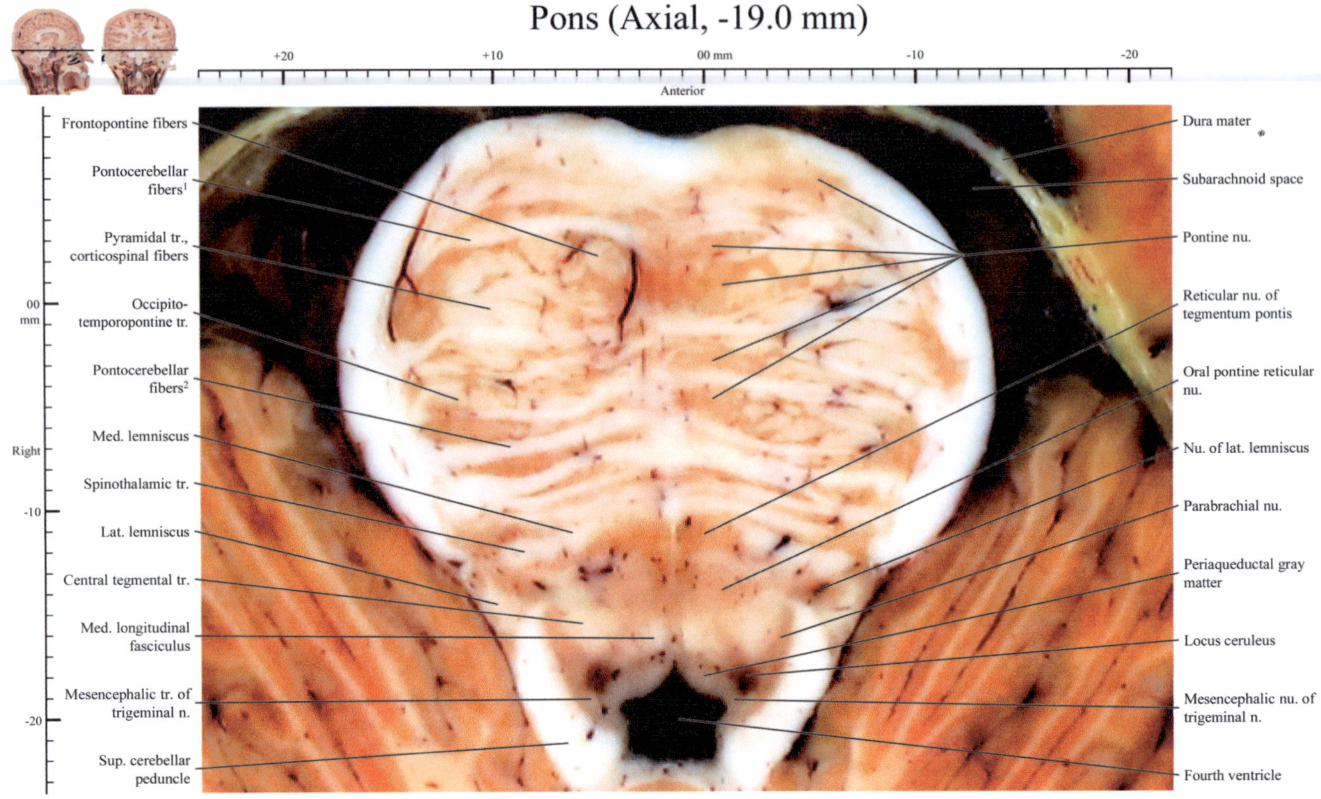

Pons (Axial, -19.4 mm)

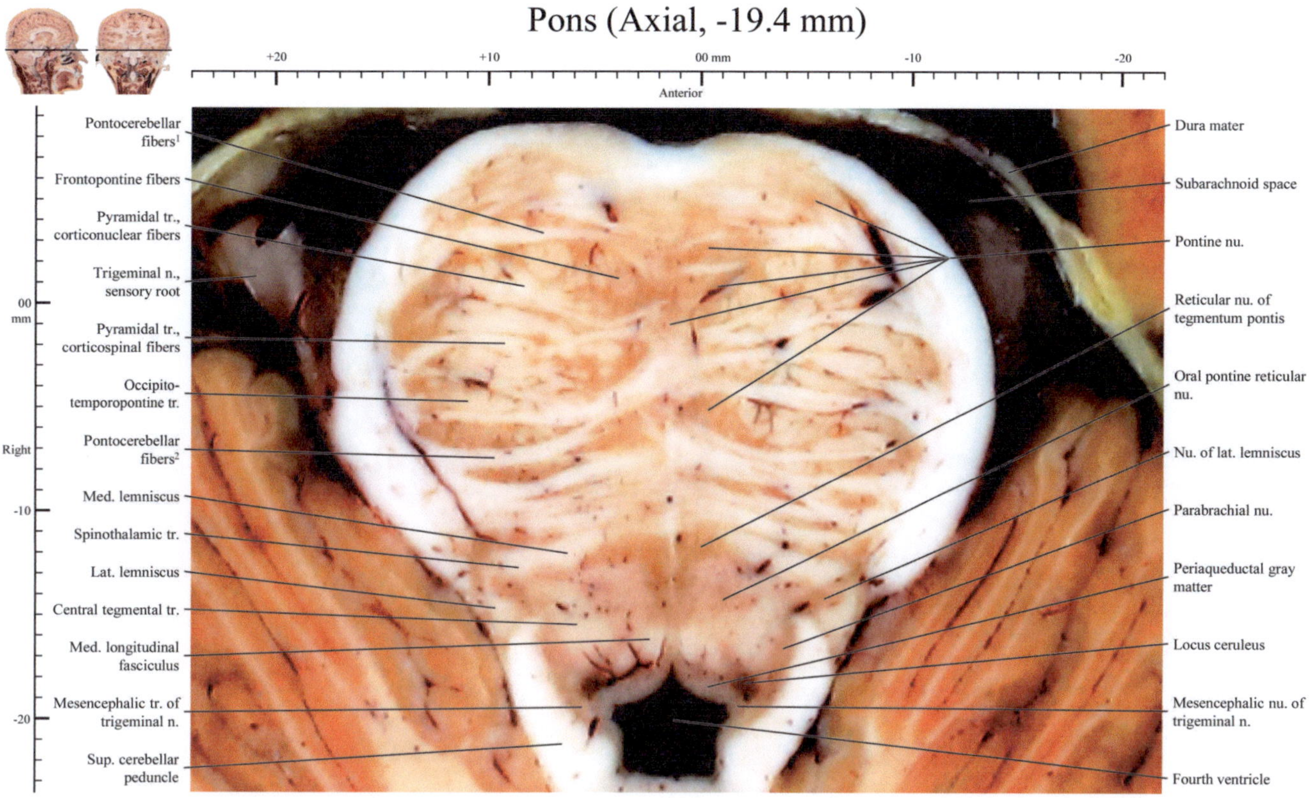

Pons (Axial, -20.0 mm)

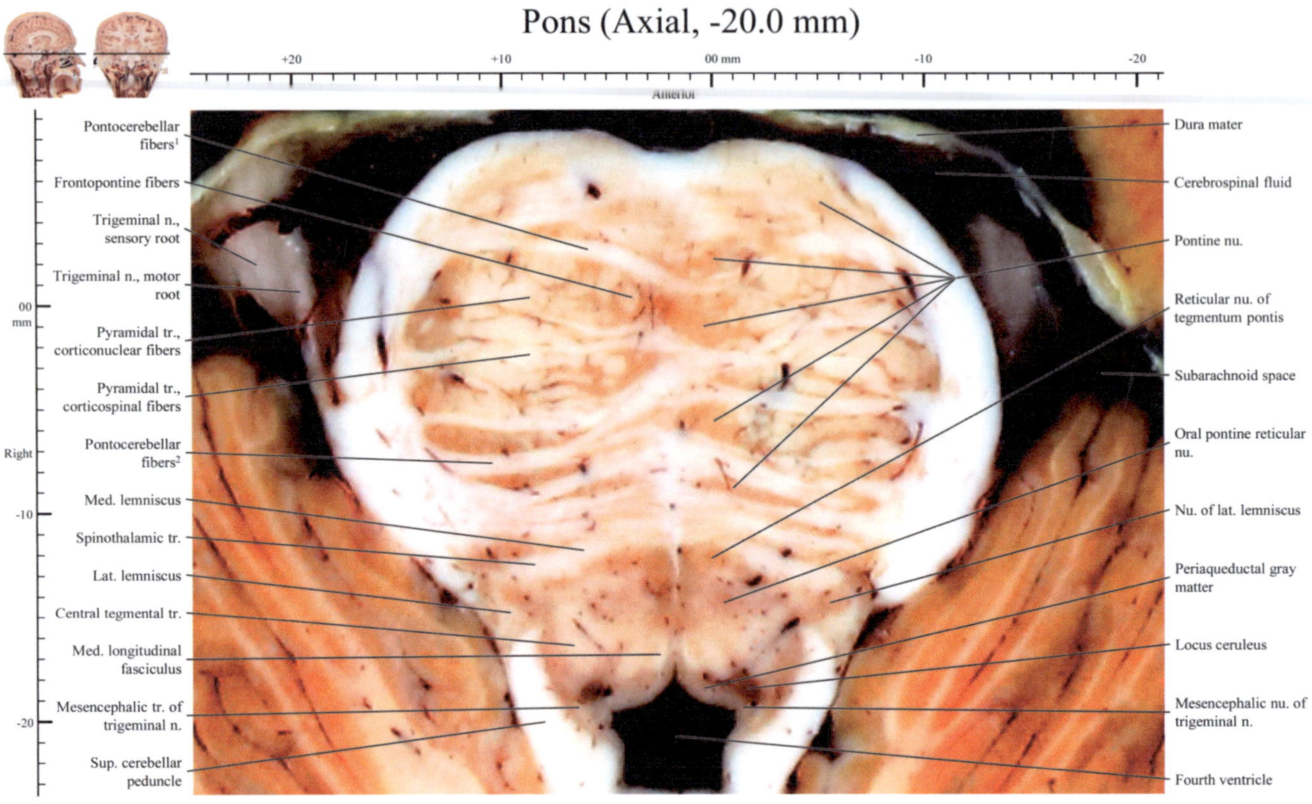

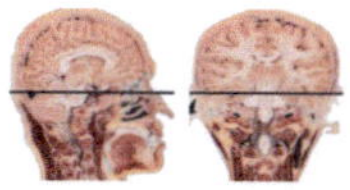

Pons (Axial, -20.4 mm)

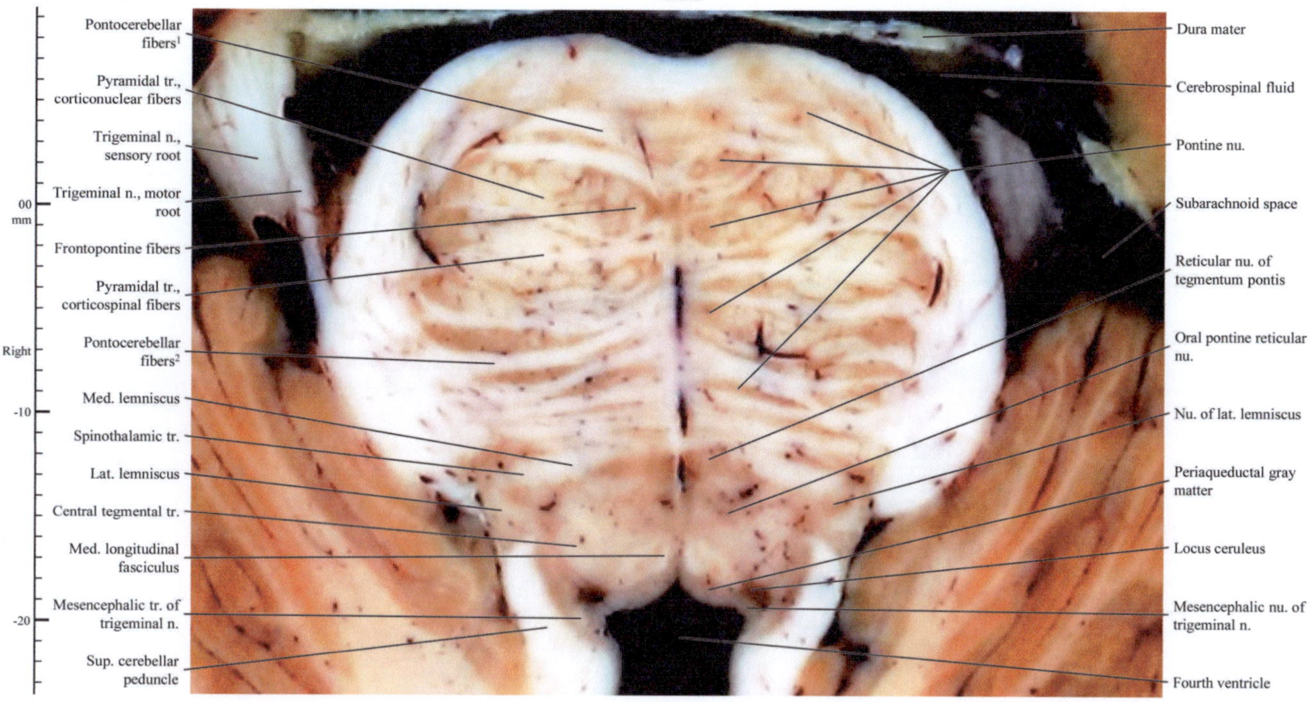

Pons (Axial, -21.0 mm)

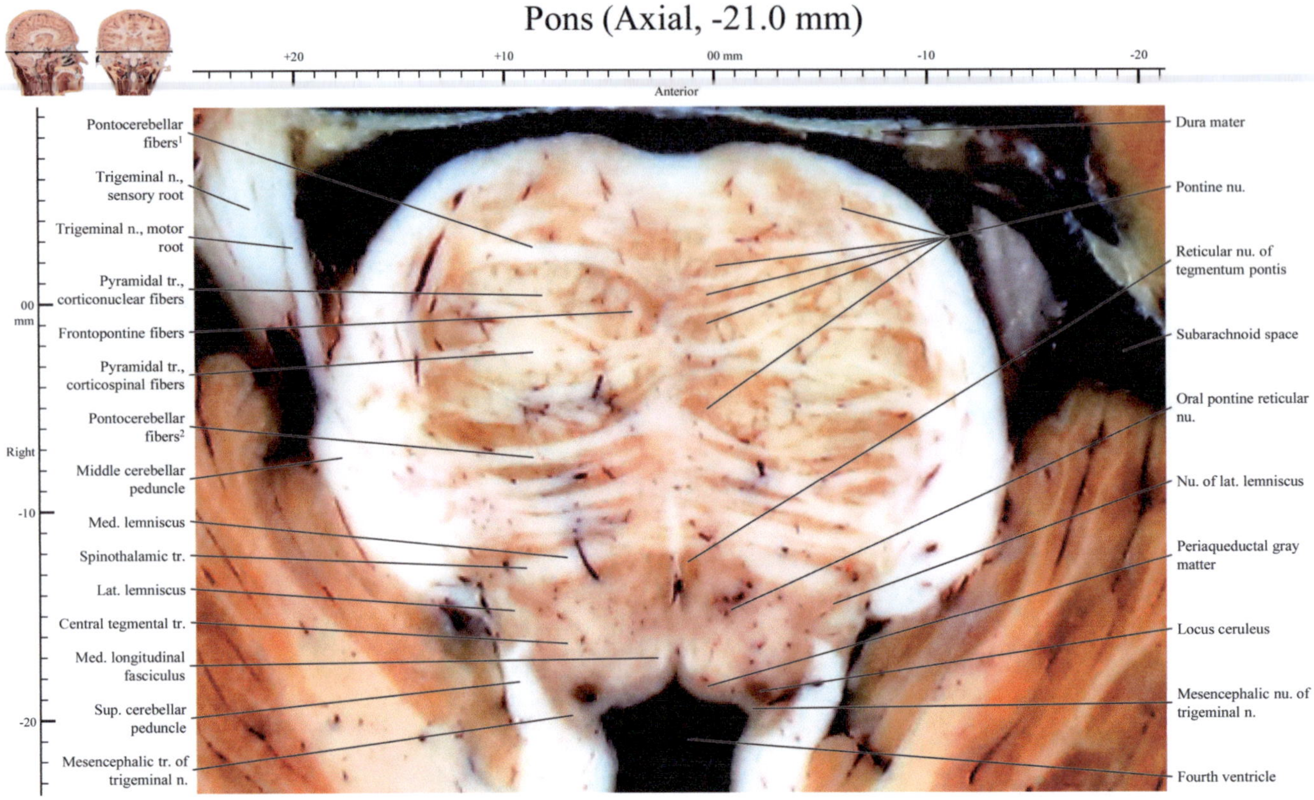

Pons (Axial, -21.4 mm)

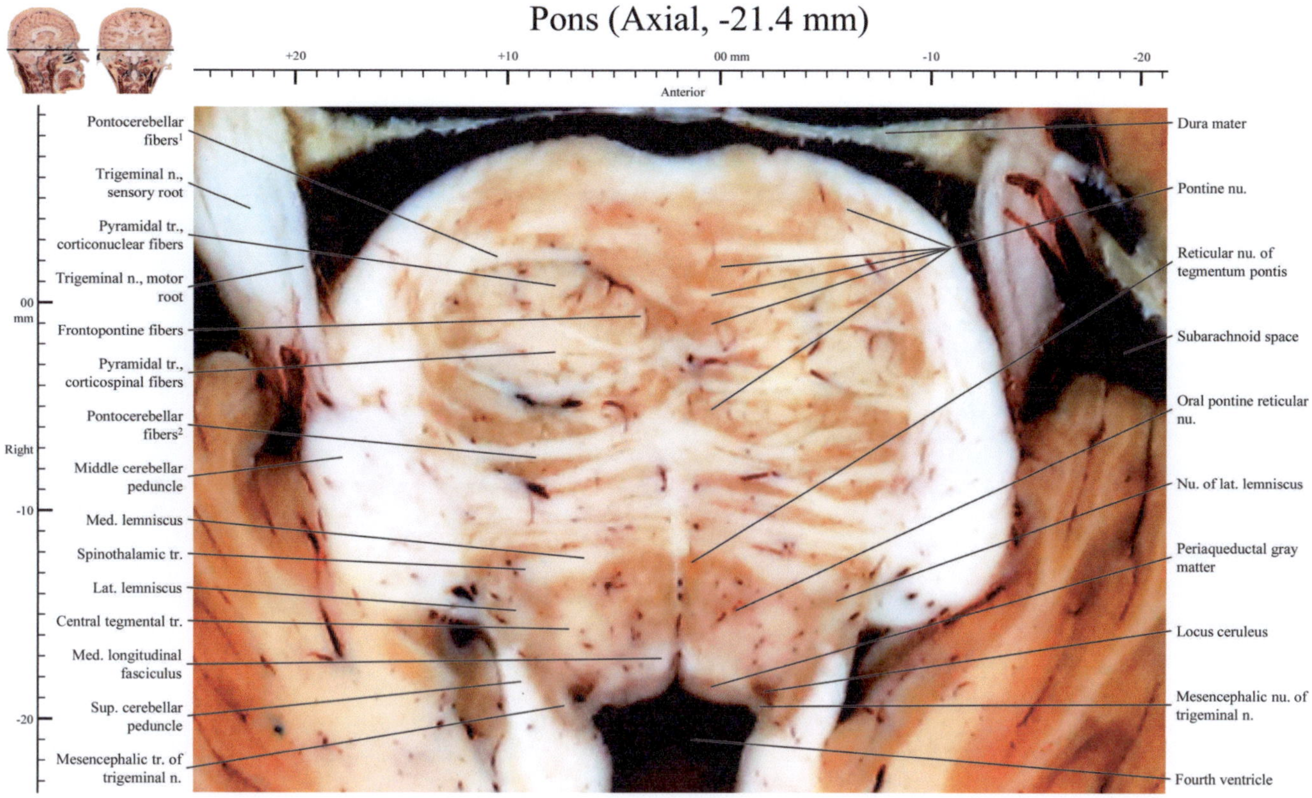

Pons (Axial, -22.0 mm)

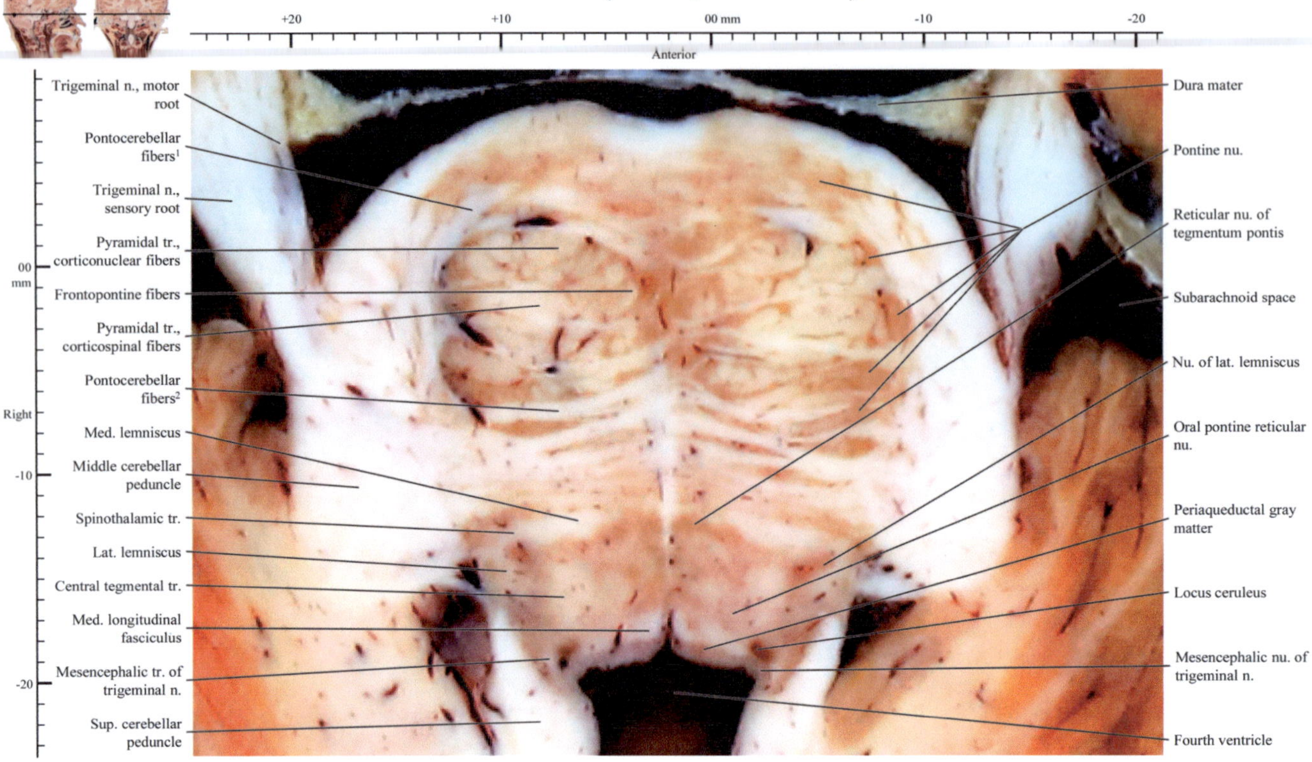

Pons (Axial, -22.4 mm)

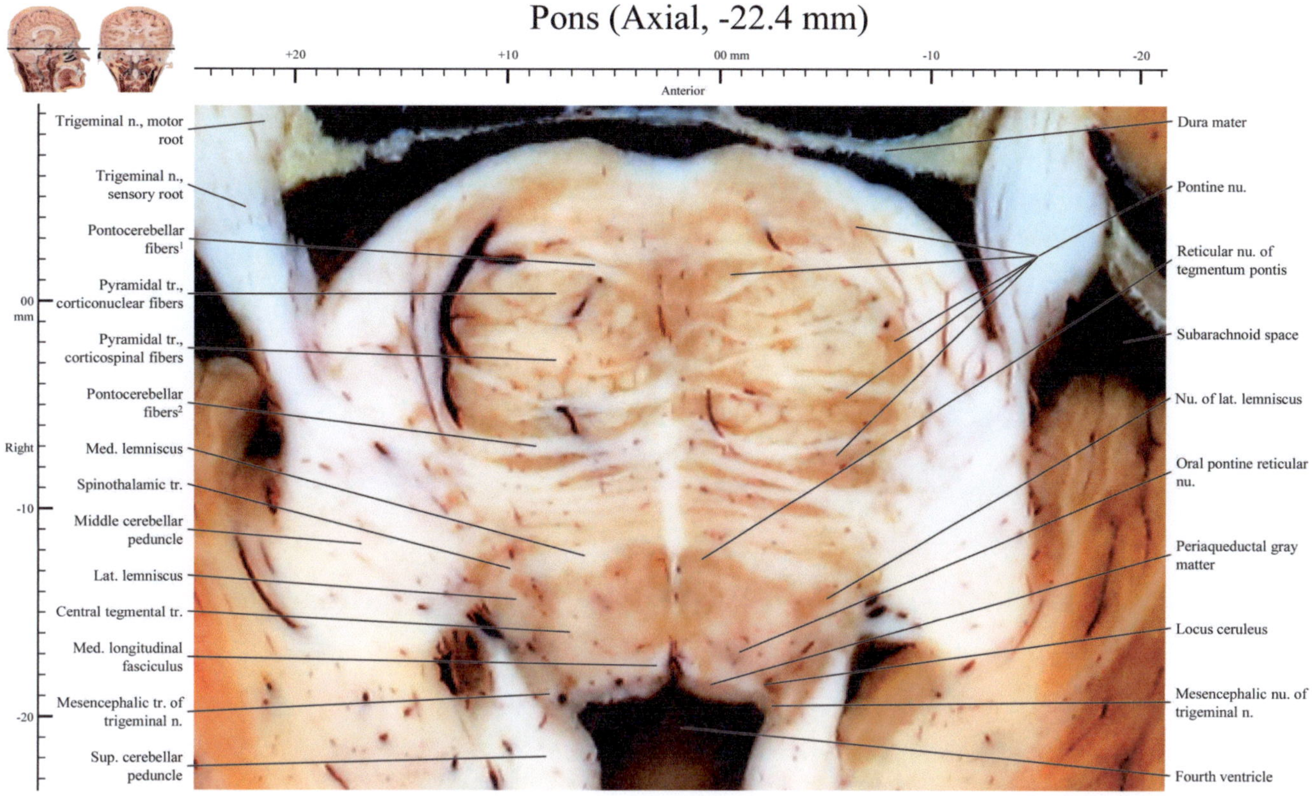

Pons (Axial, -23.0 mm)

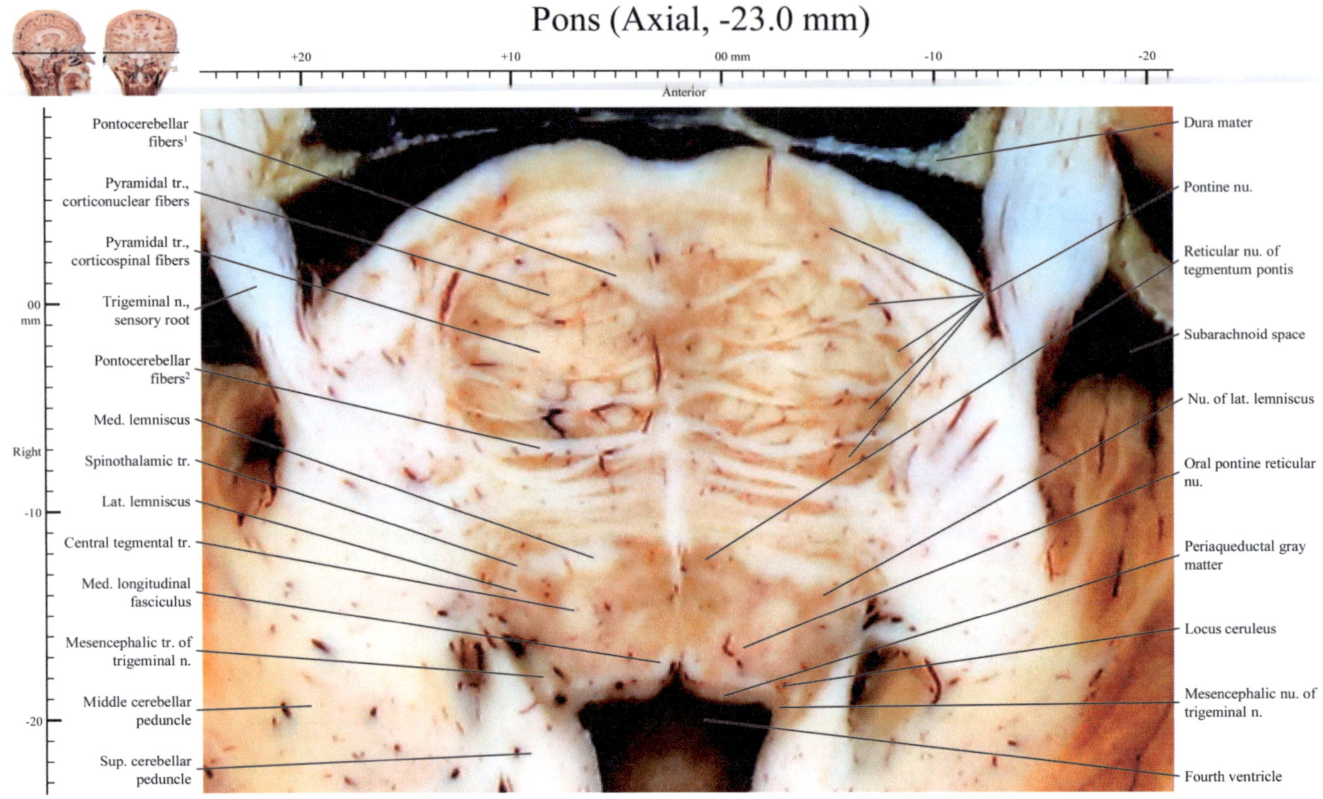

Pons (Axial, -23.4 mm)

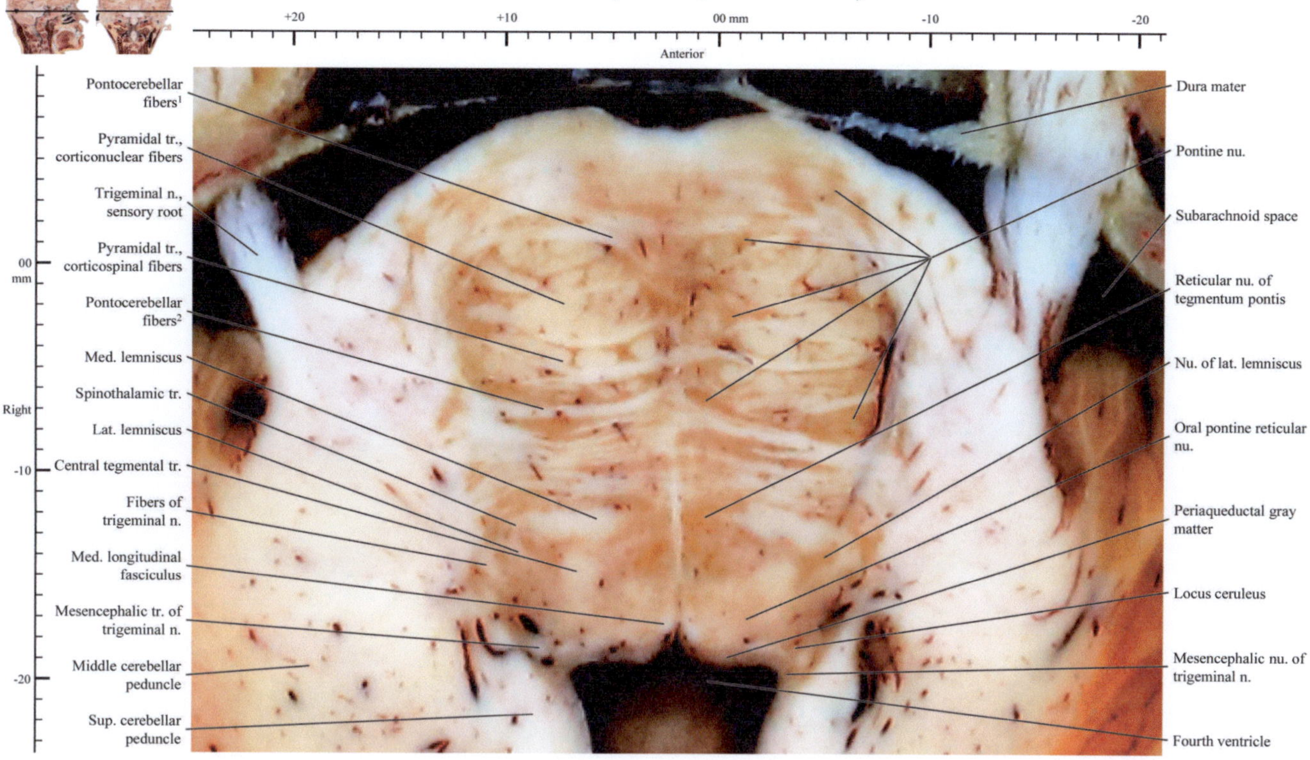

Pons (Axial, -24.0 mm)

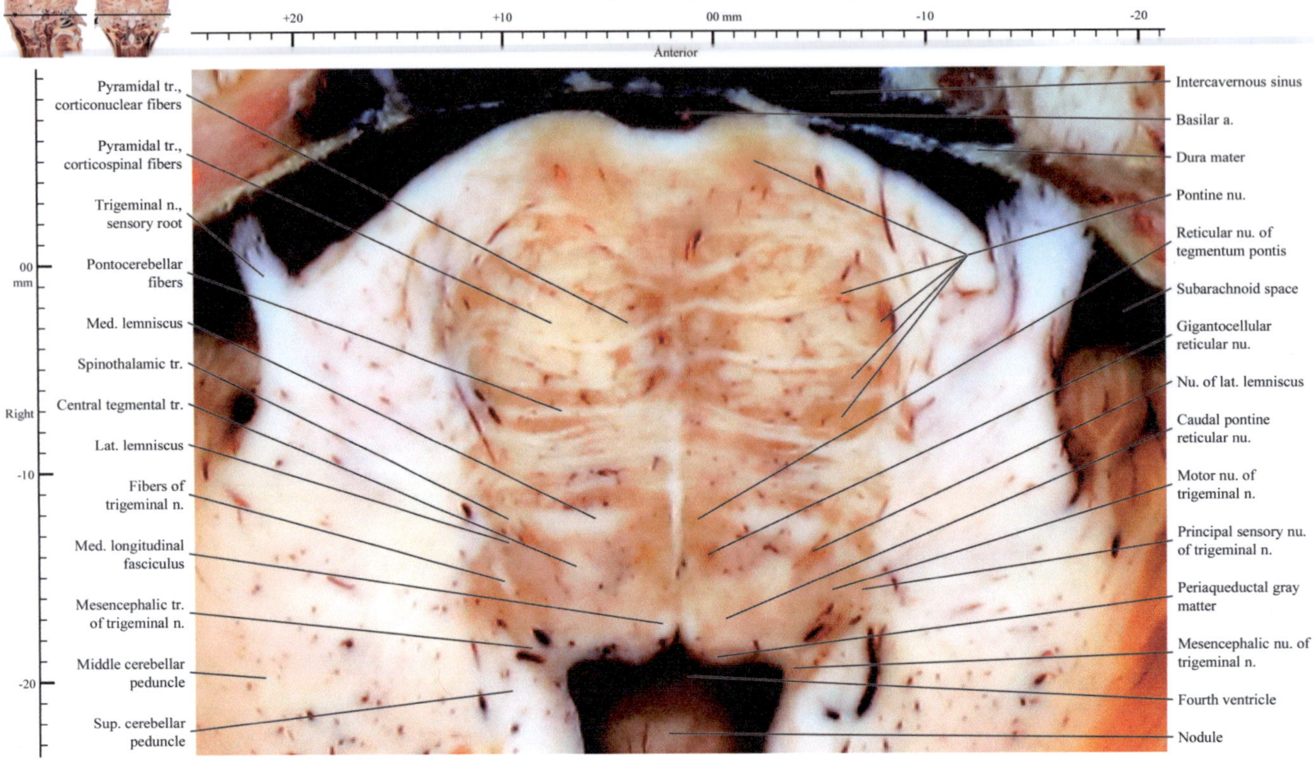

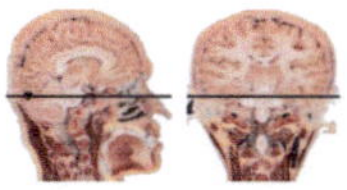

Pons (Axial, -24.4 mm)

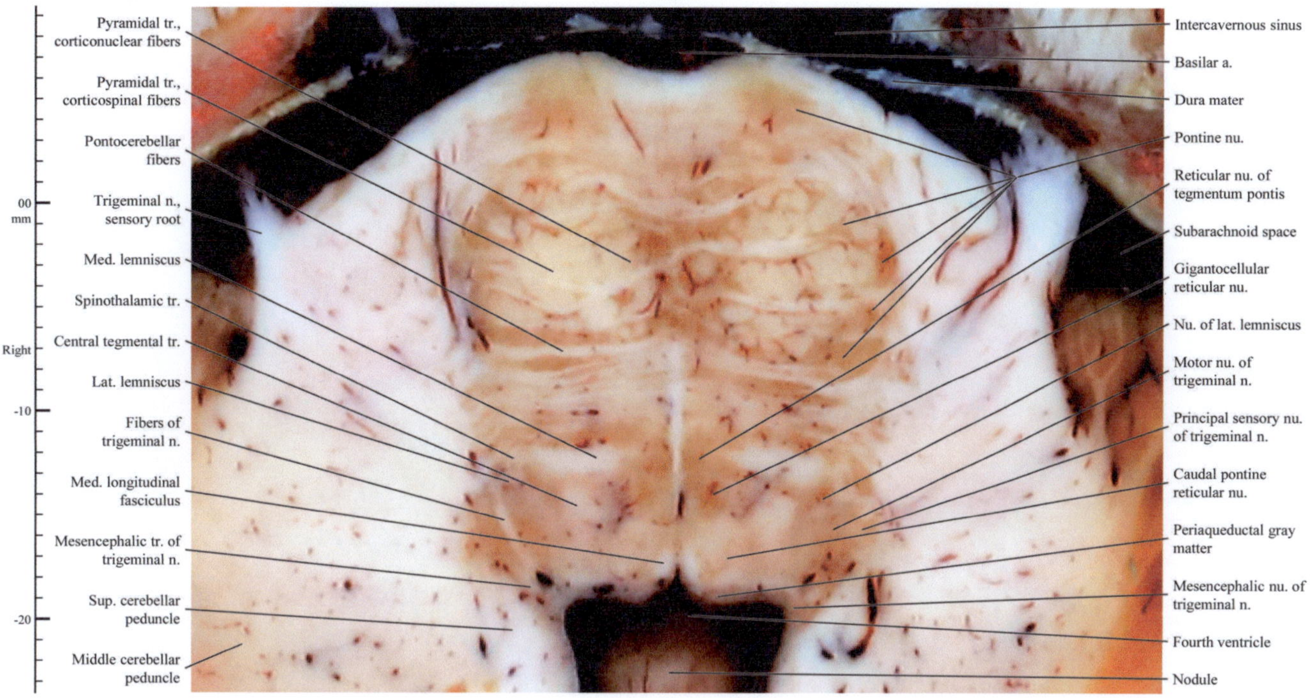

Pons (Axial, -25.0 mm)

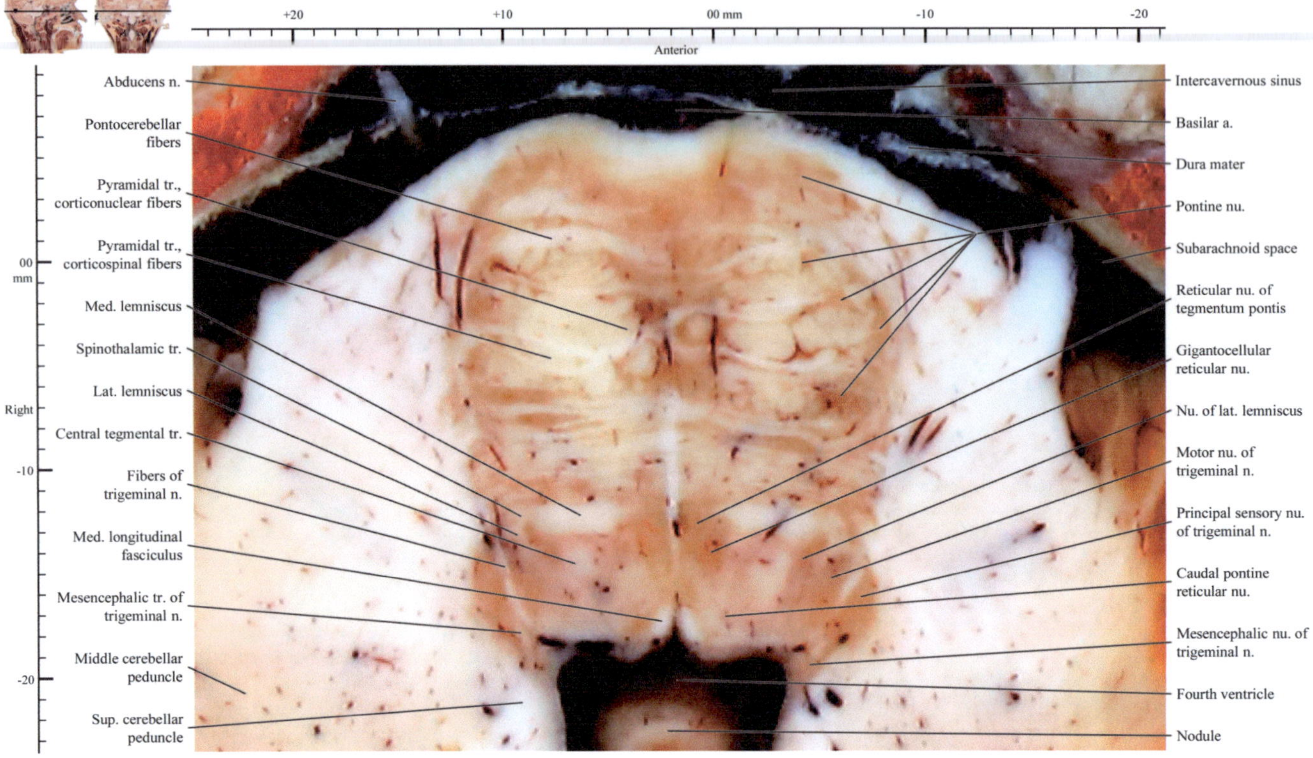

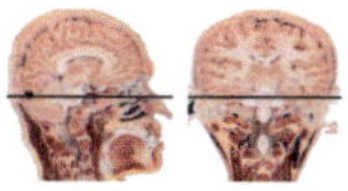

Pons (Axial, -25.4 mm)

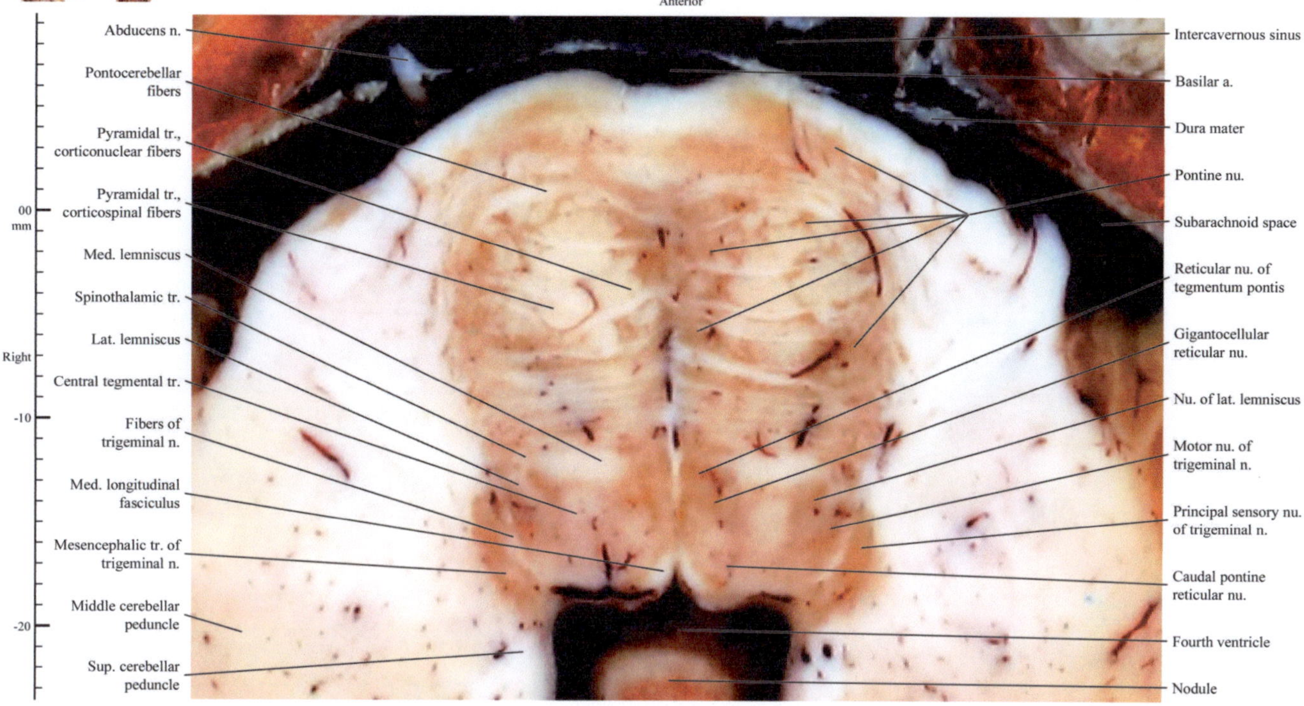

Pons (Axial, -26.0 mm)

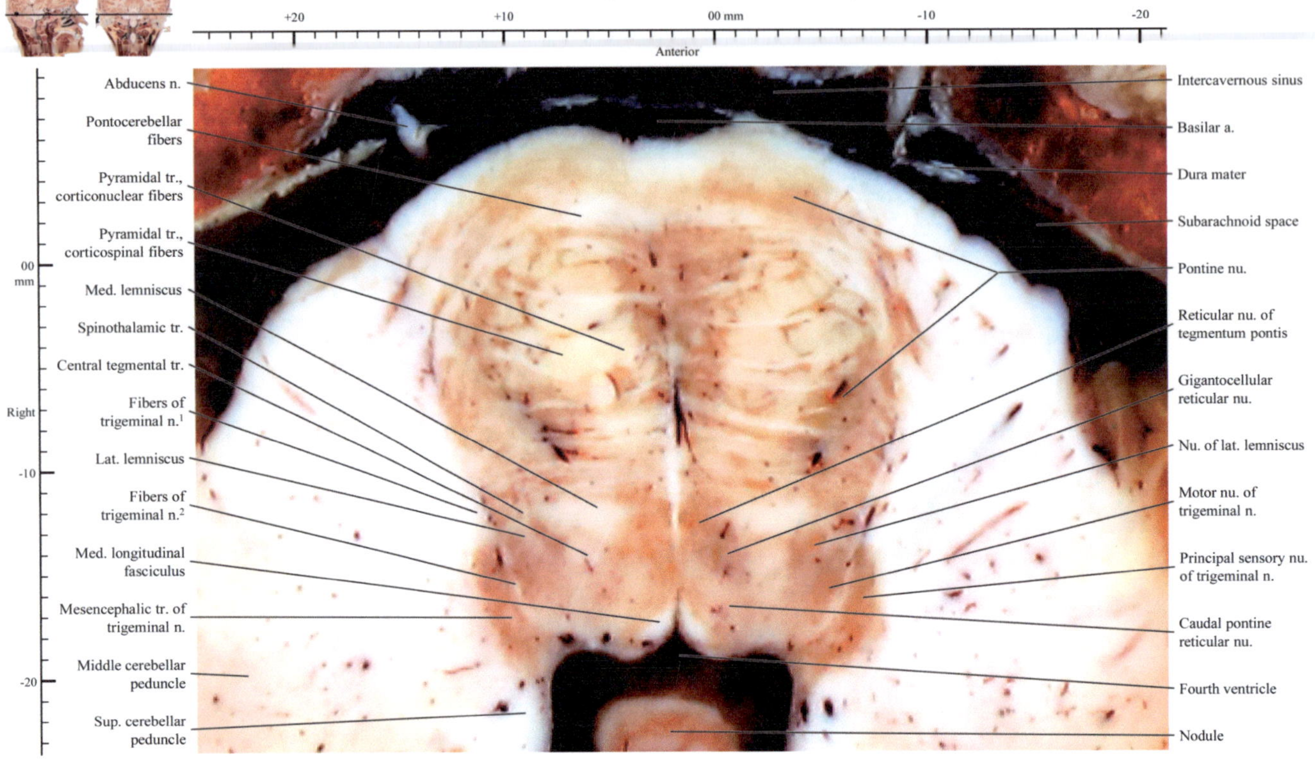

Pons (Axial, -26.4 mm)

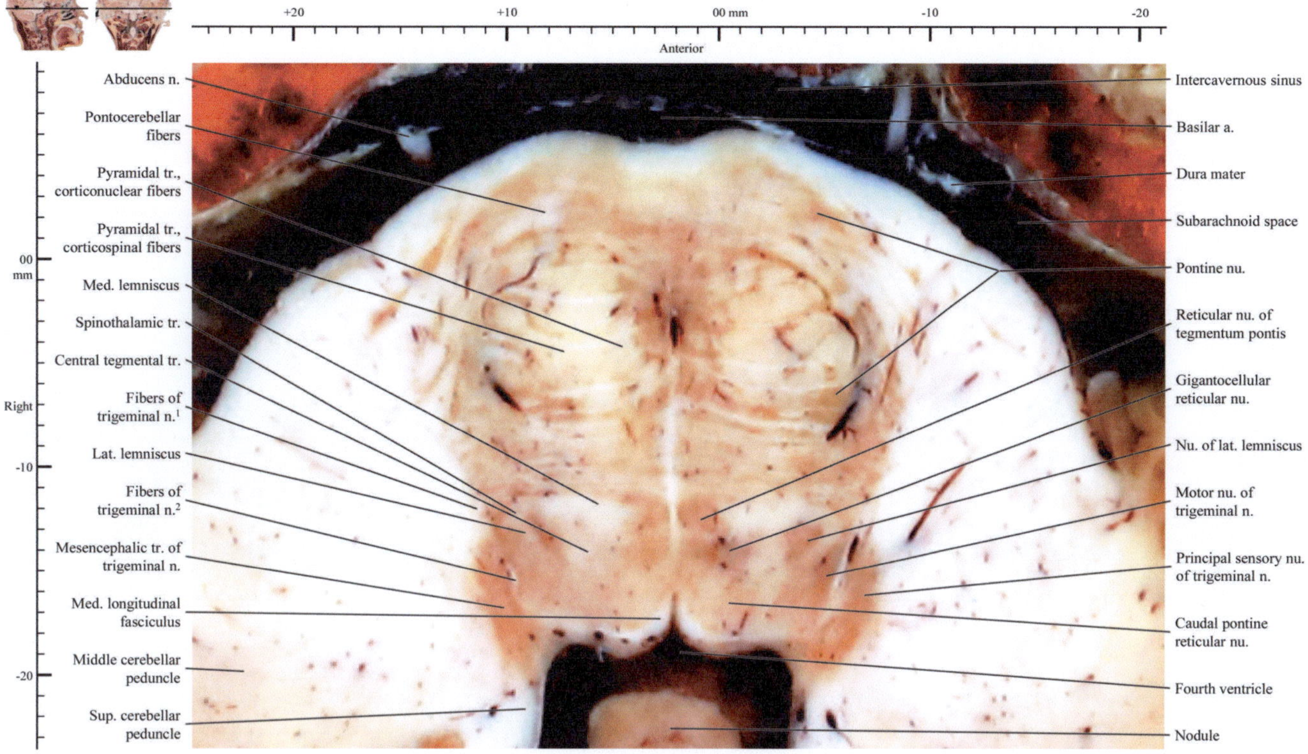

Pons (Axial, -27.0 mm)

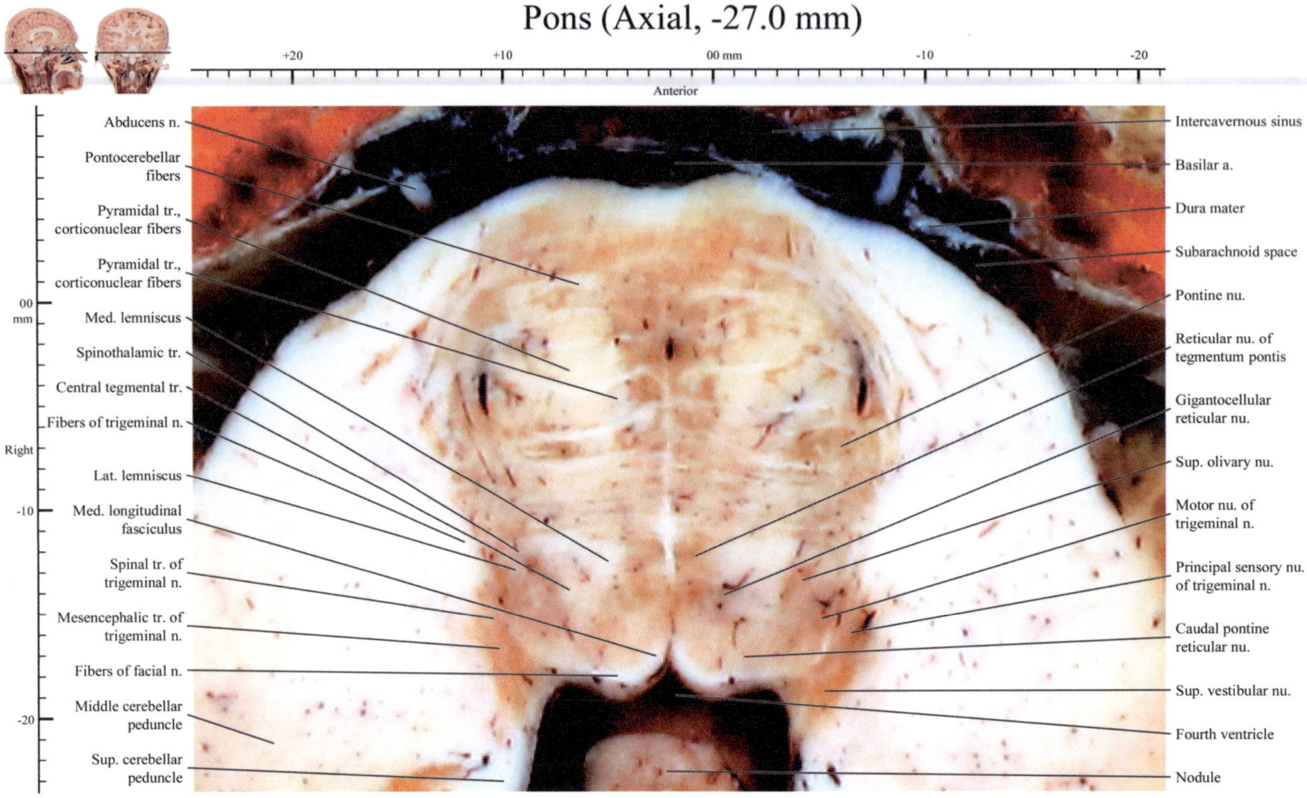

Pons (Axial, -27.4 mm)

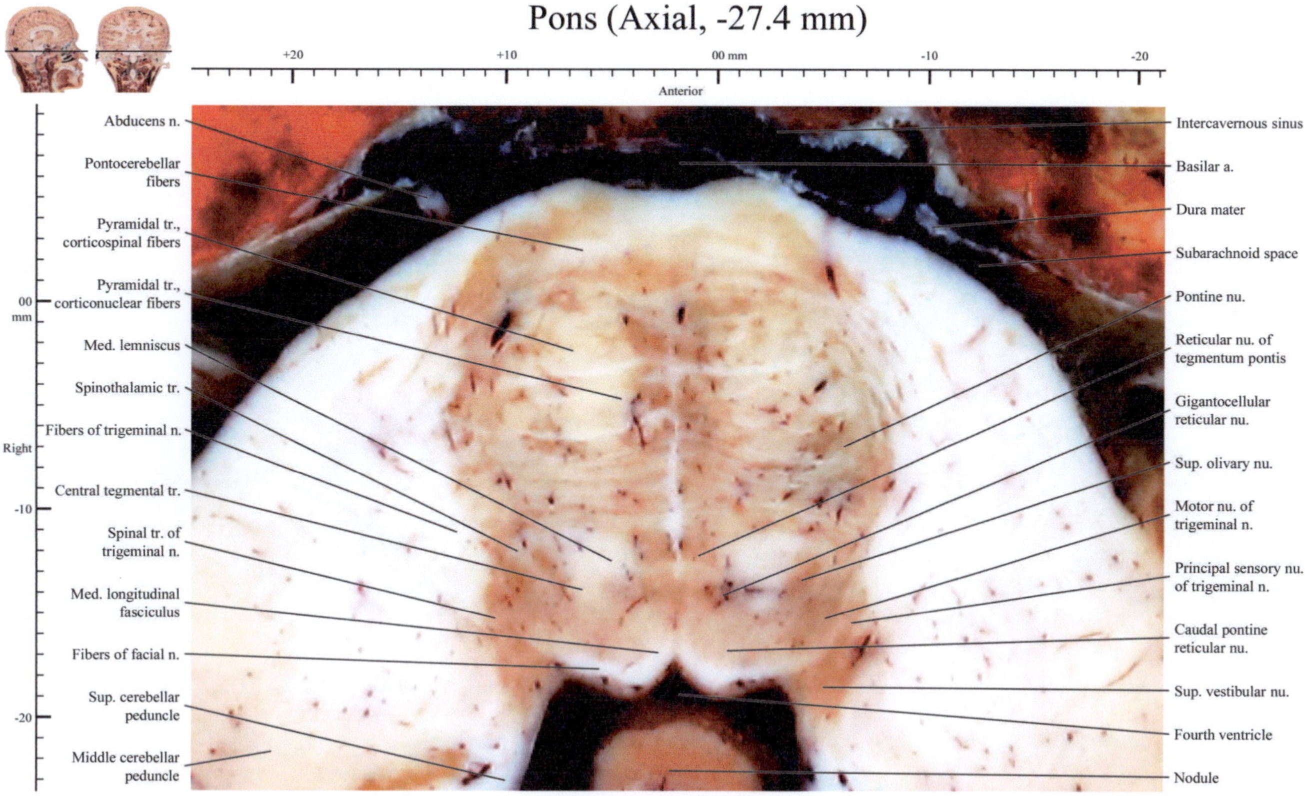

Pons (Axial, -28.0 mm)

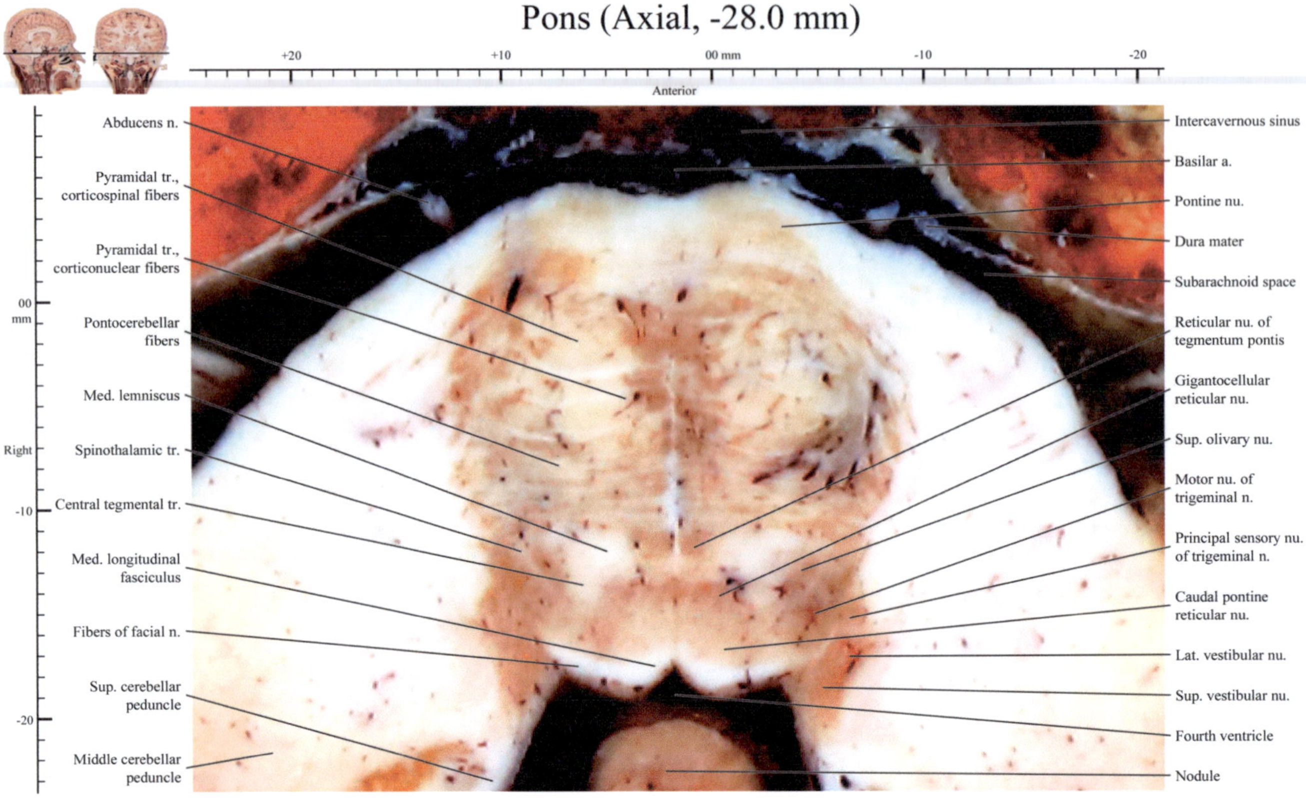

Pons (Axial, -28.4 mm)

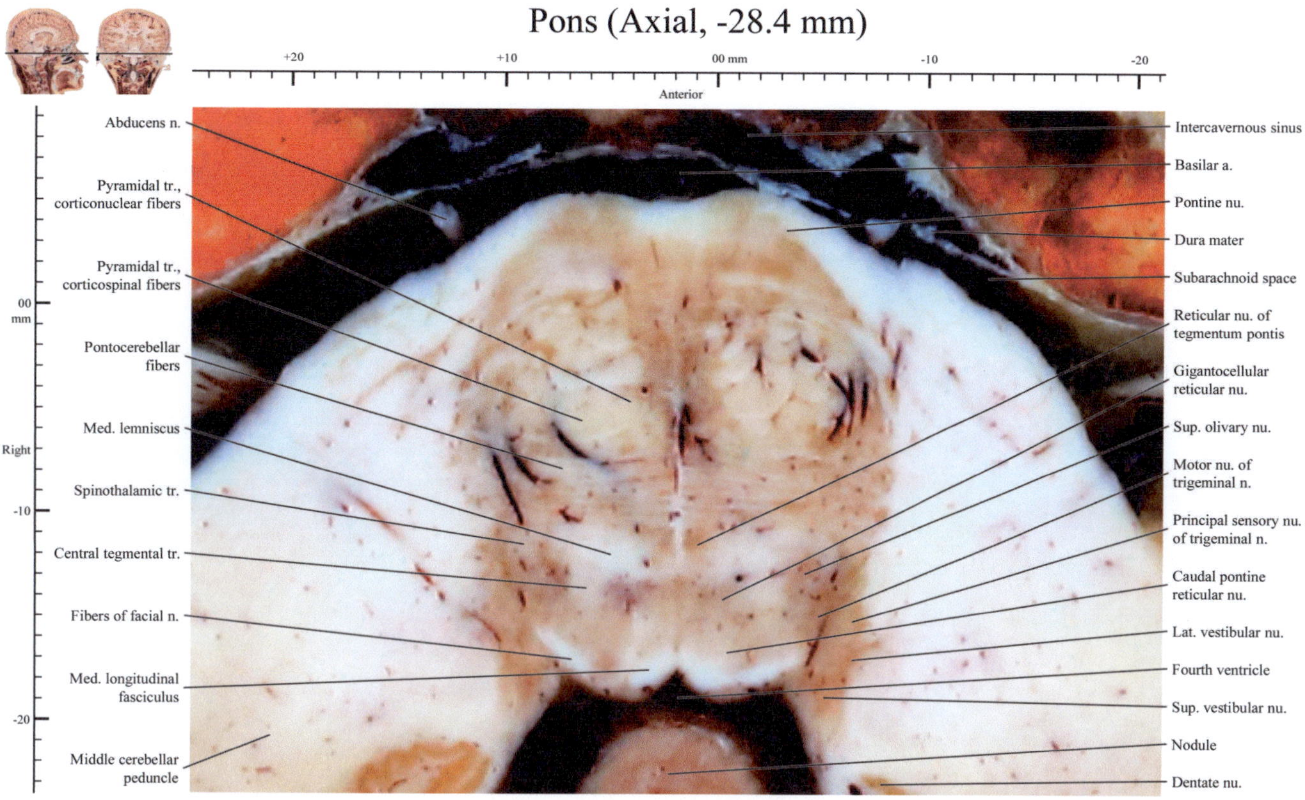

Pons (Axial, -29.0 mm)

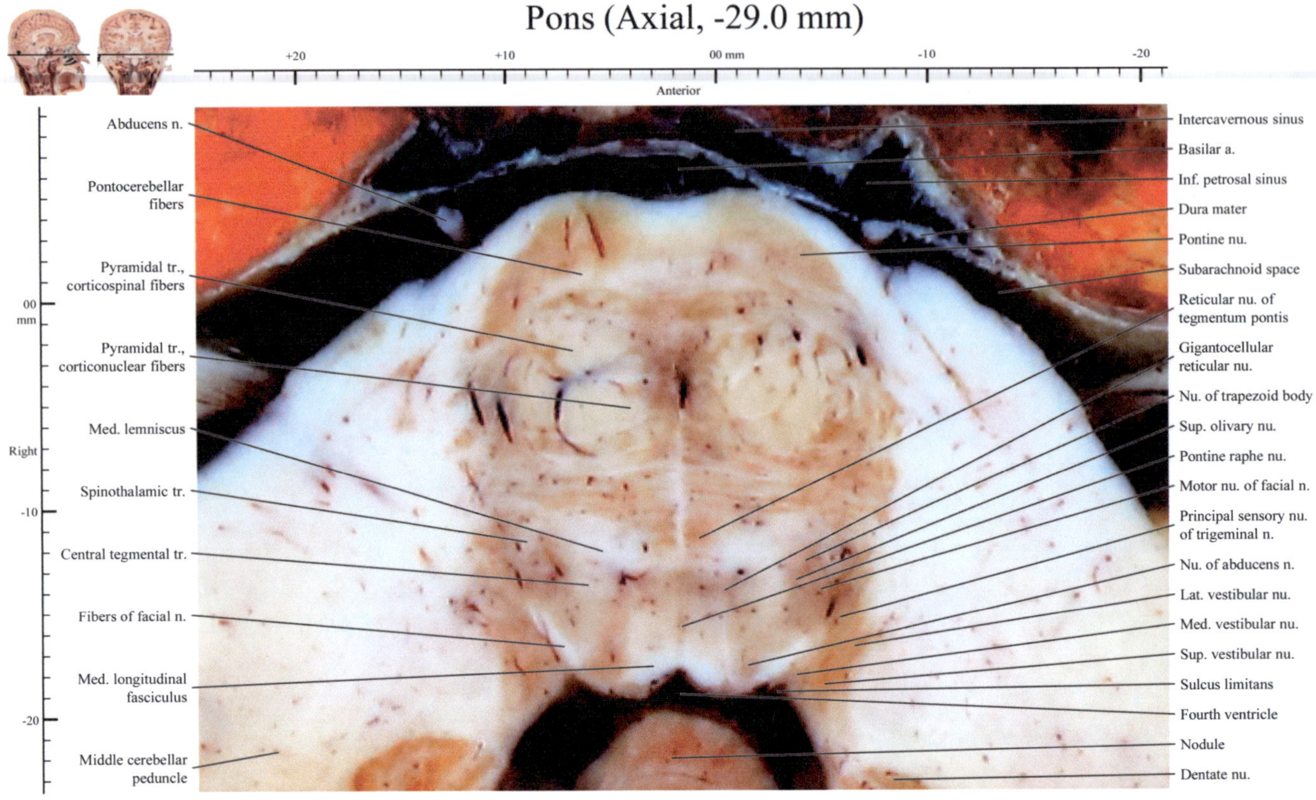

Pons (Axial, -29.4 mm)

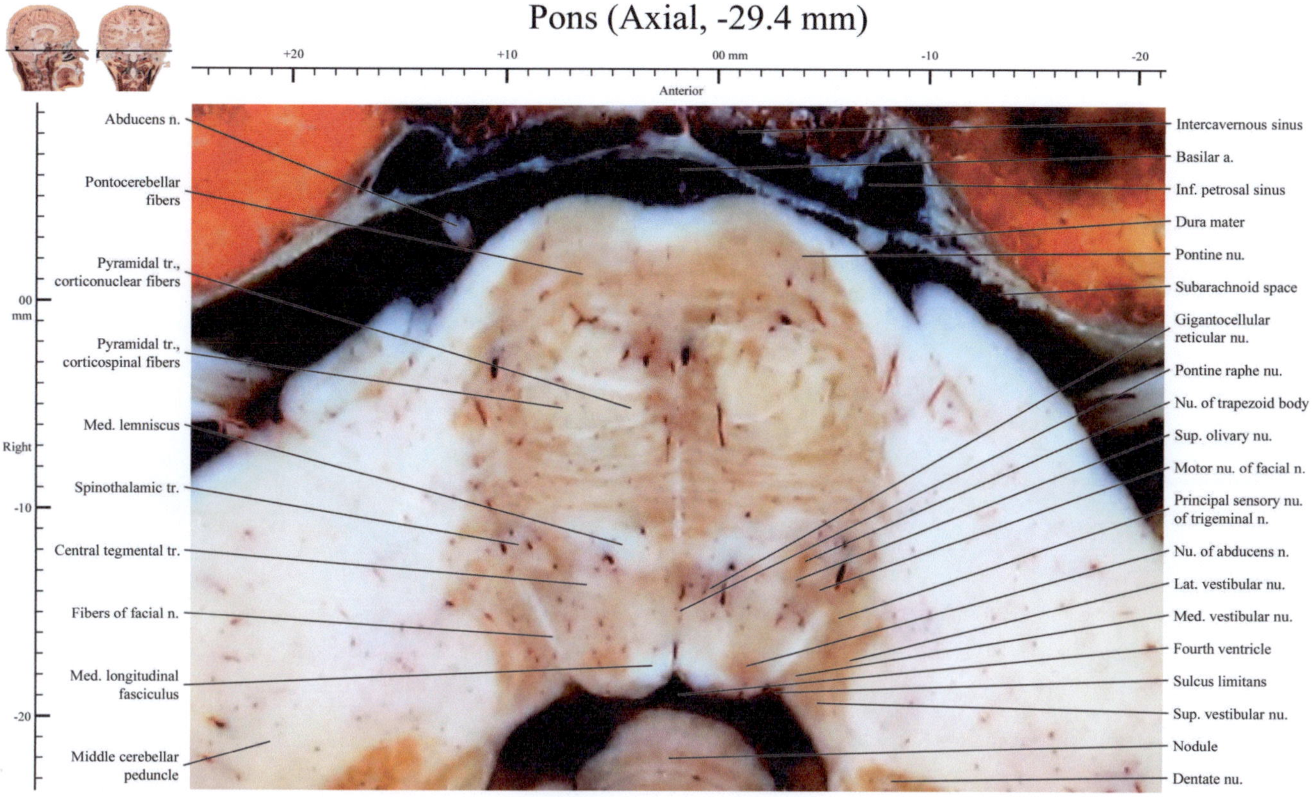

Pons (Axial, -30.0 mm)

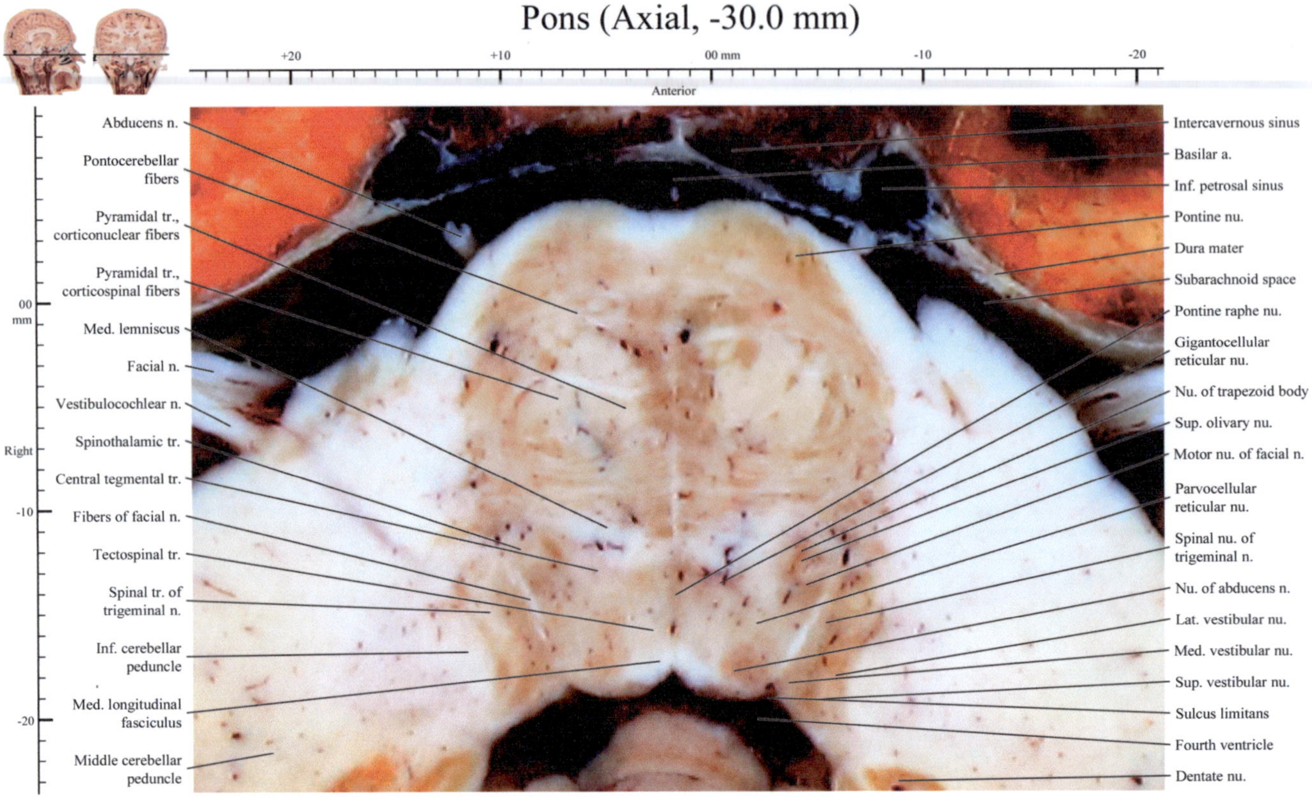

Pons (Axial, -30.4 mm)

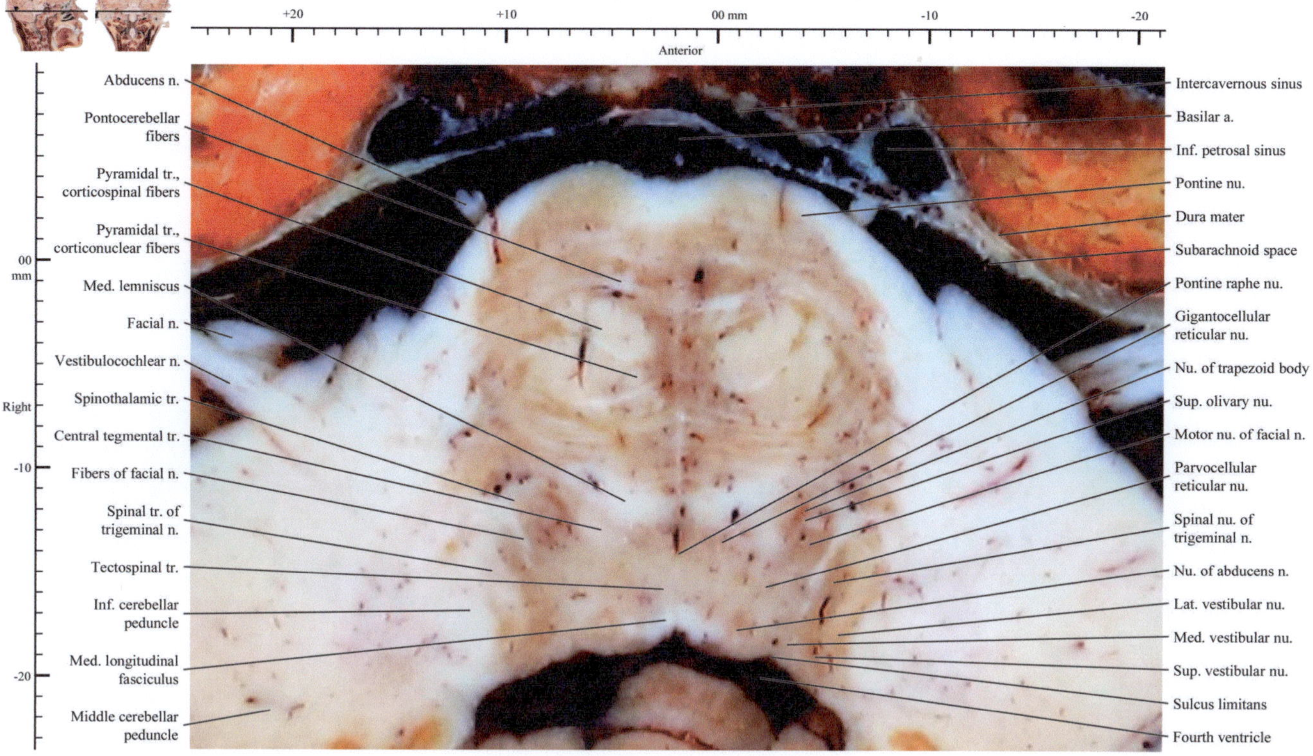

Pons (Axial, -31.0 mm)

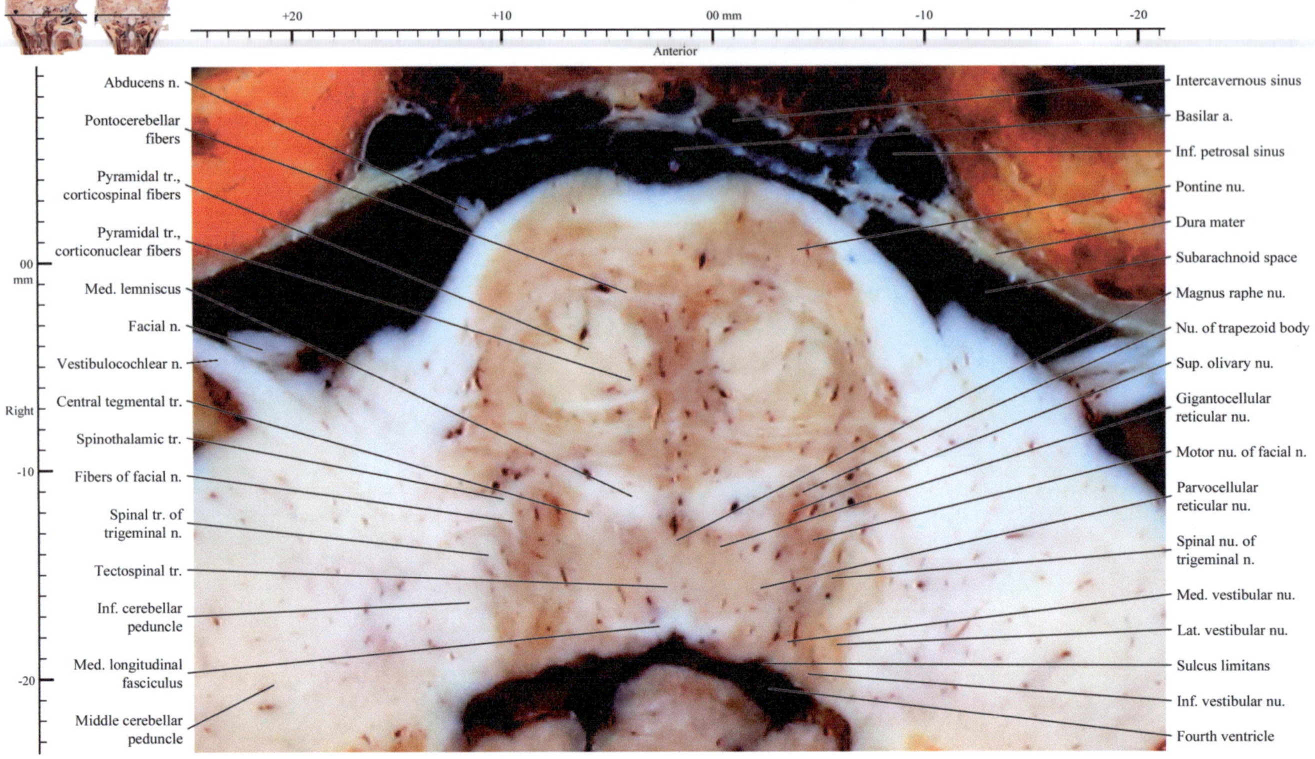

Pons (Axial, -31.4 mm)

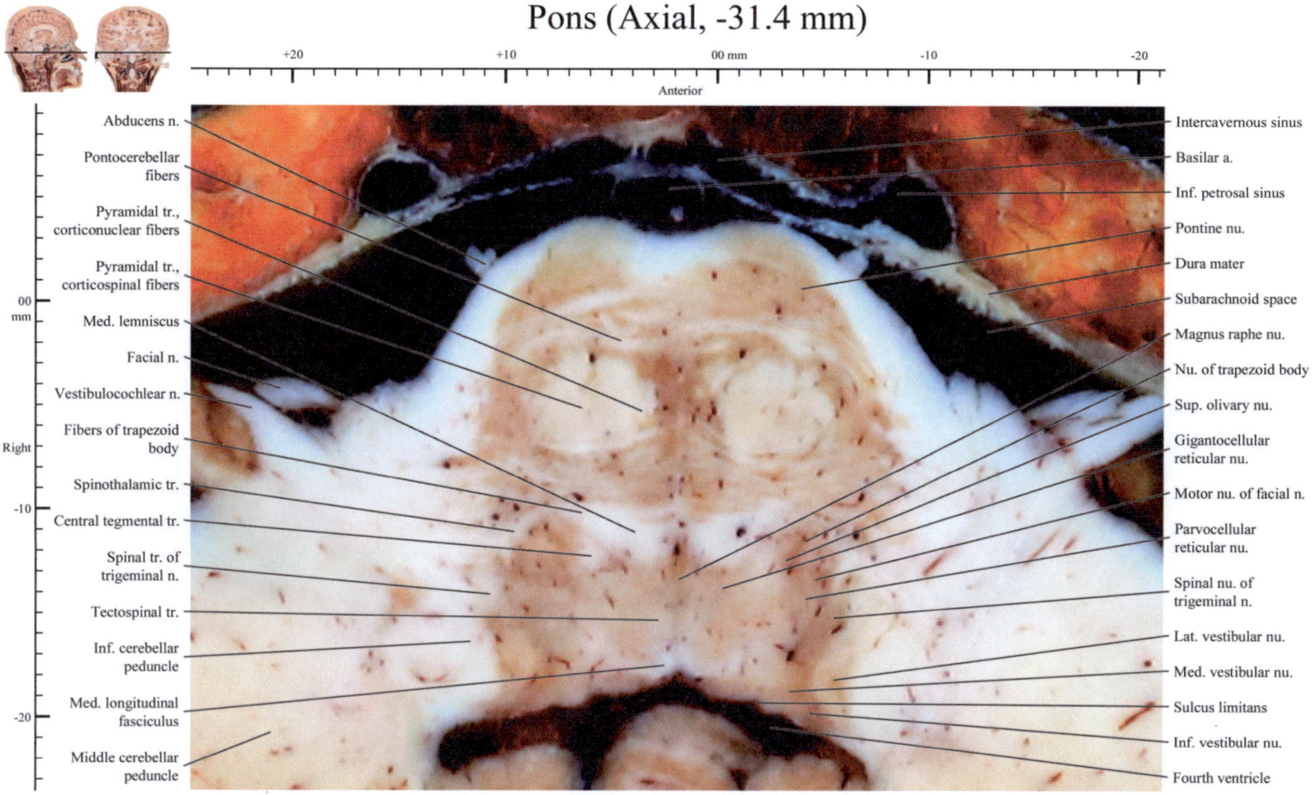

Pons (Axial, -32.0 mm)

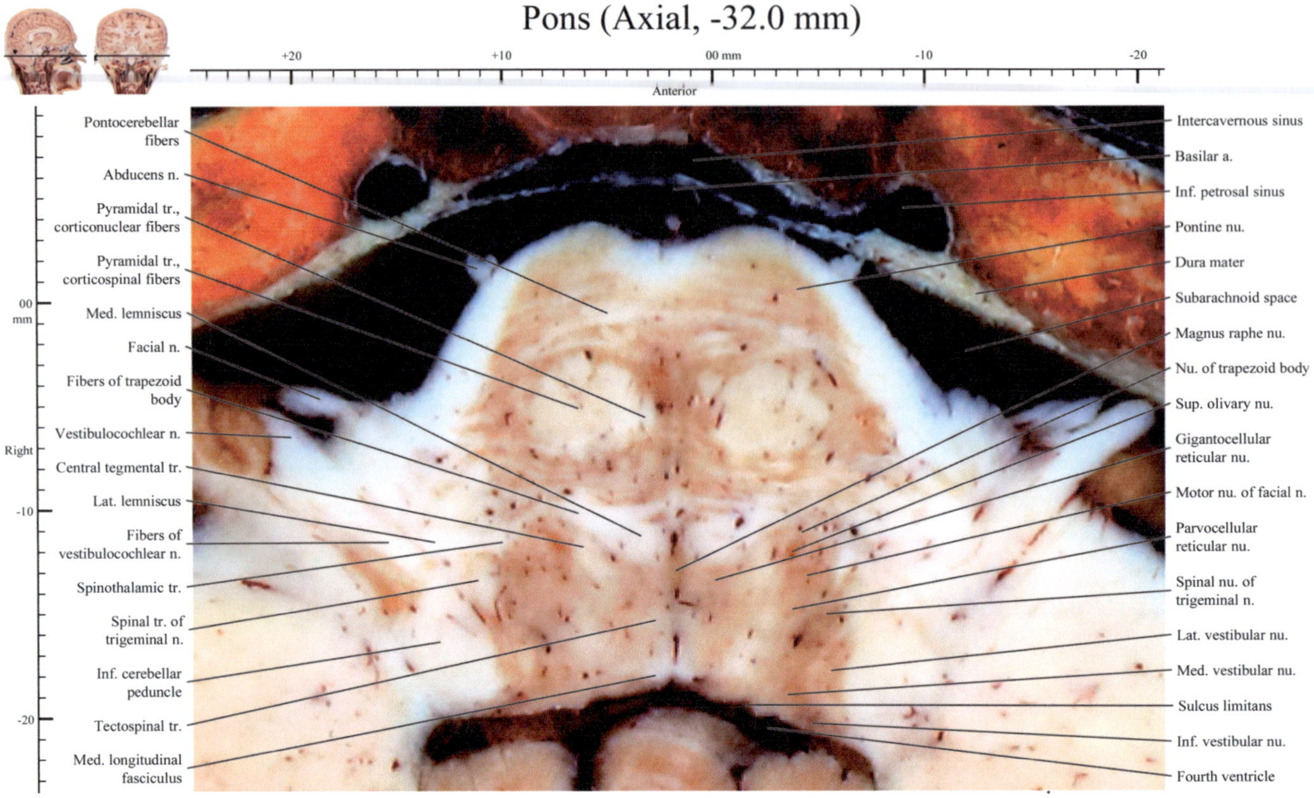

Pons (Axial, -32.4 mm)

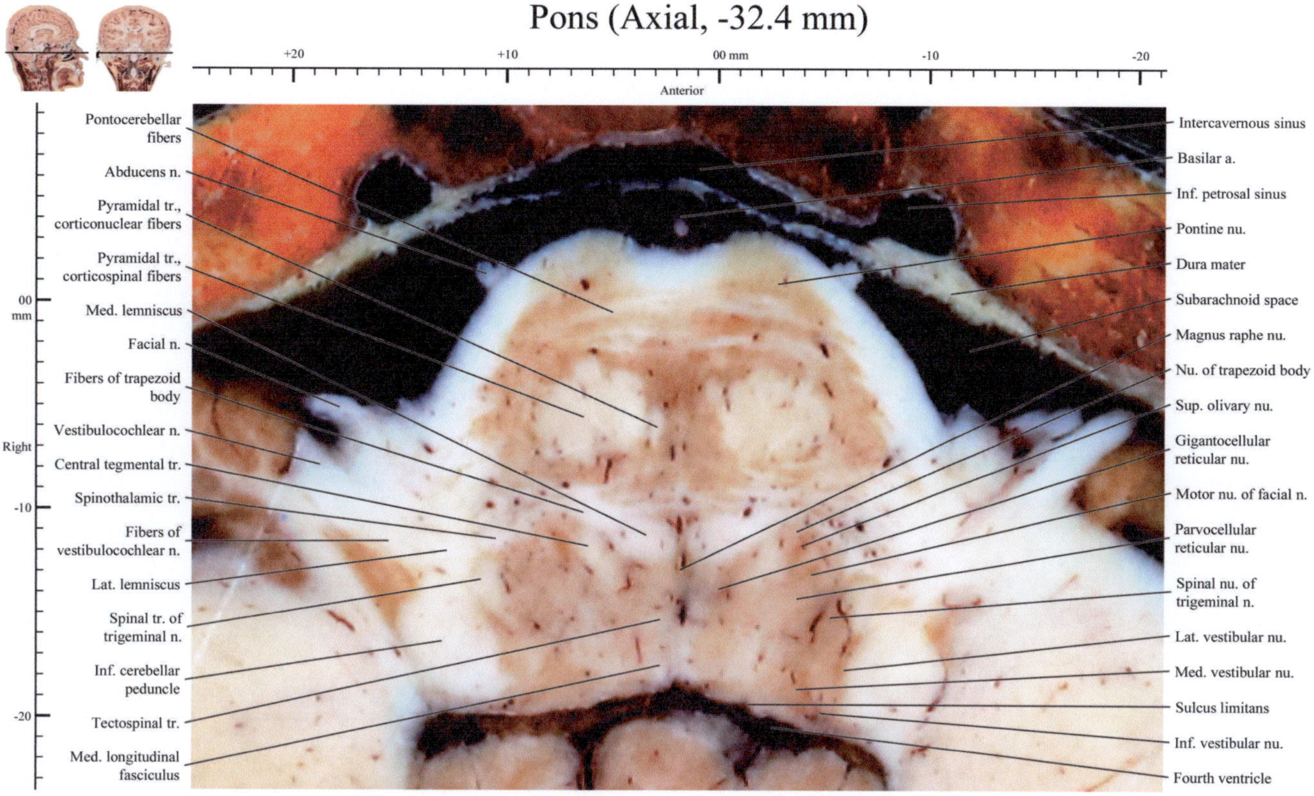

Pons and medulla oblongata (Axial, -33.0 mm)

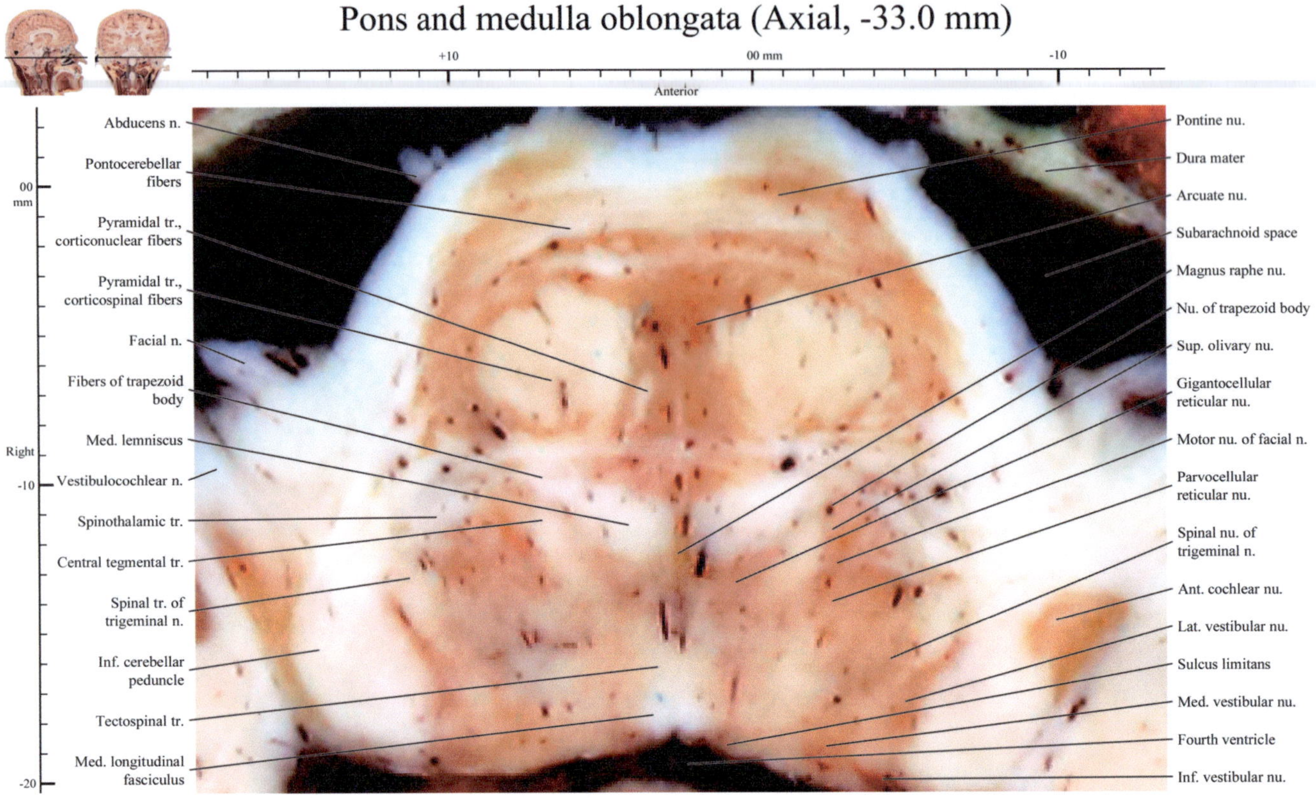

Pons and medulla oblongata (Axial, -33.4 mm)

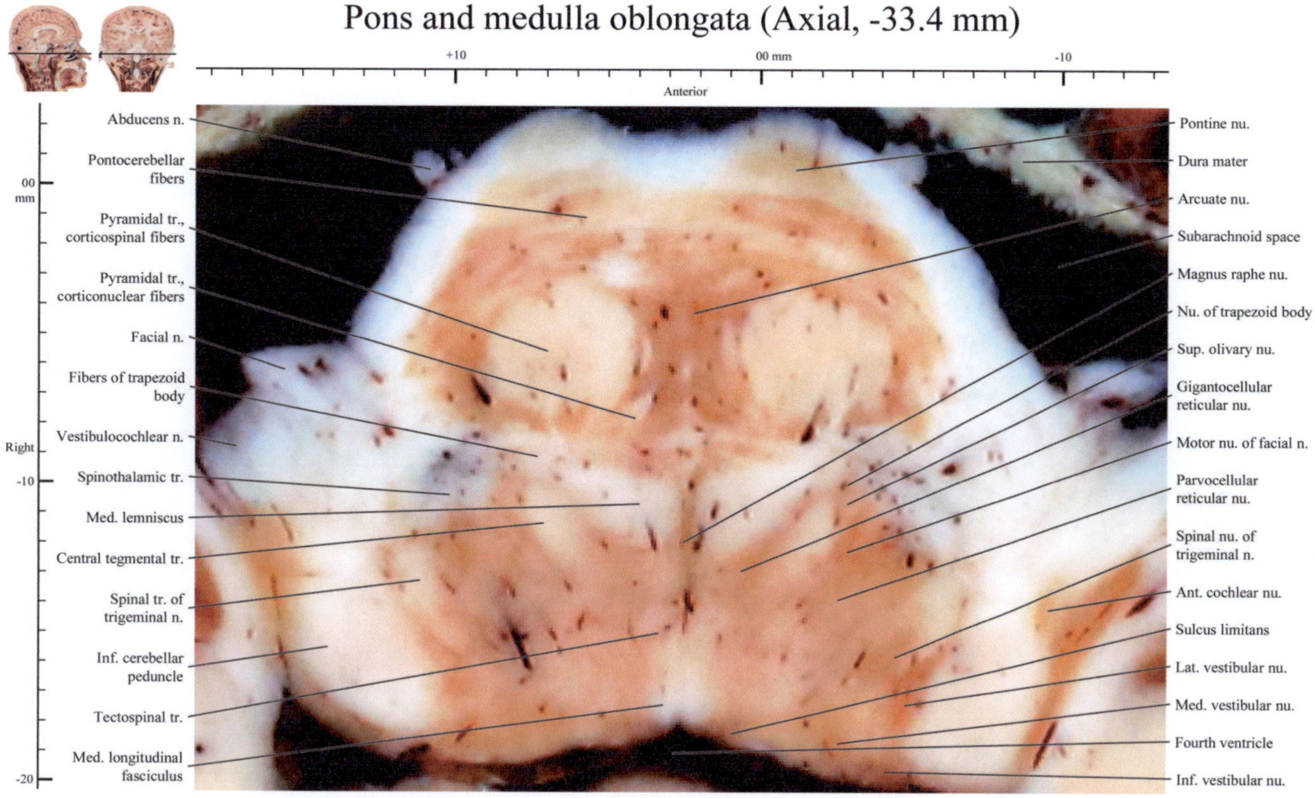

Pons and medulla oblongata (Axial, -34.0 mm)

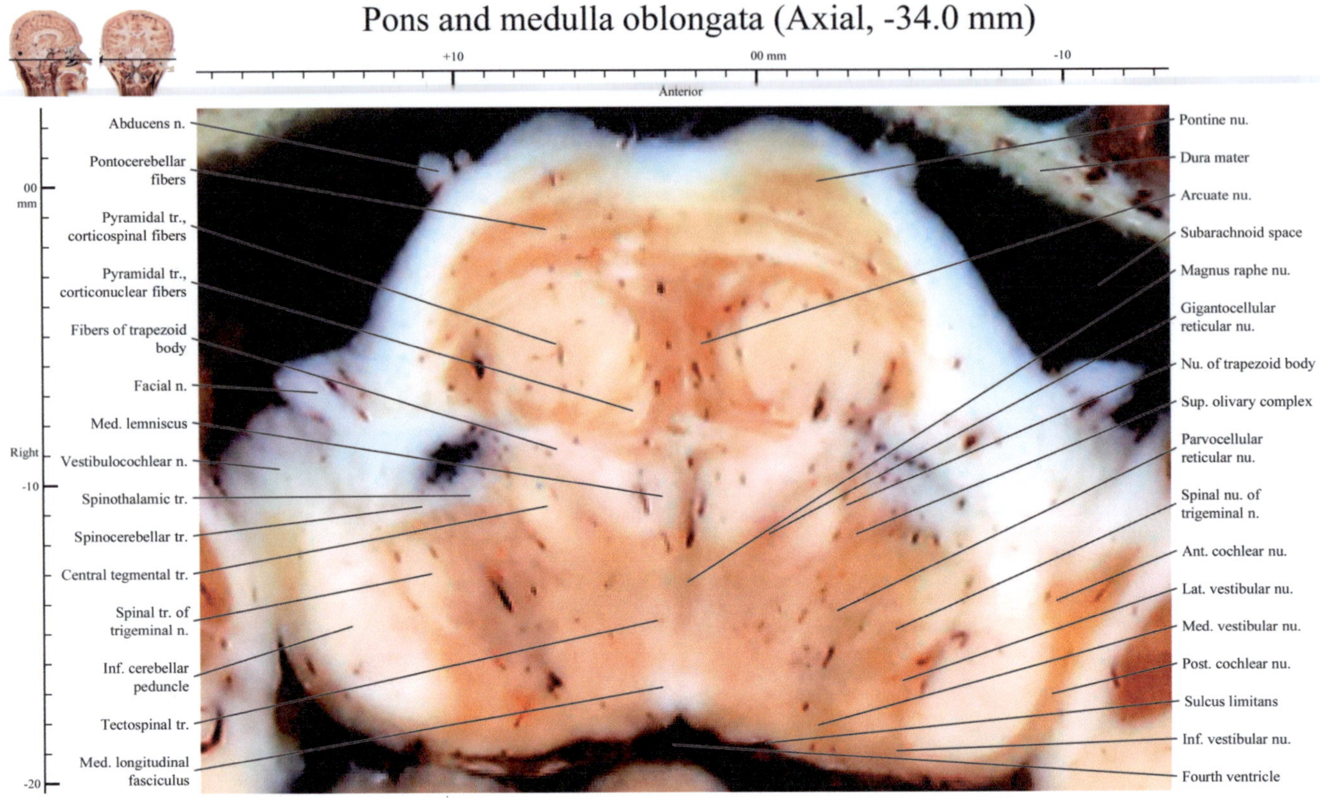

Pons and medulla oblongata (Axial, -34.4 mm)

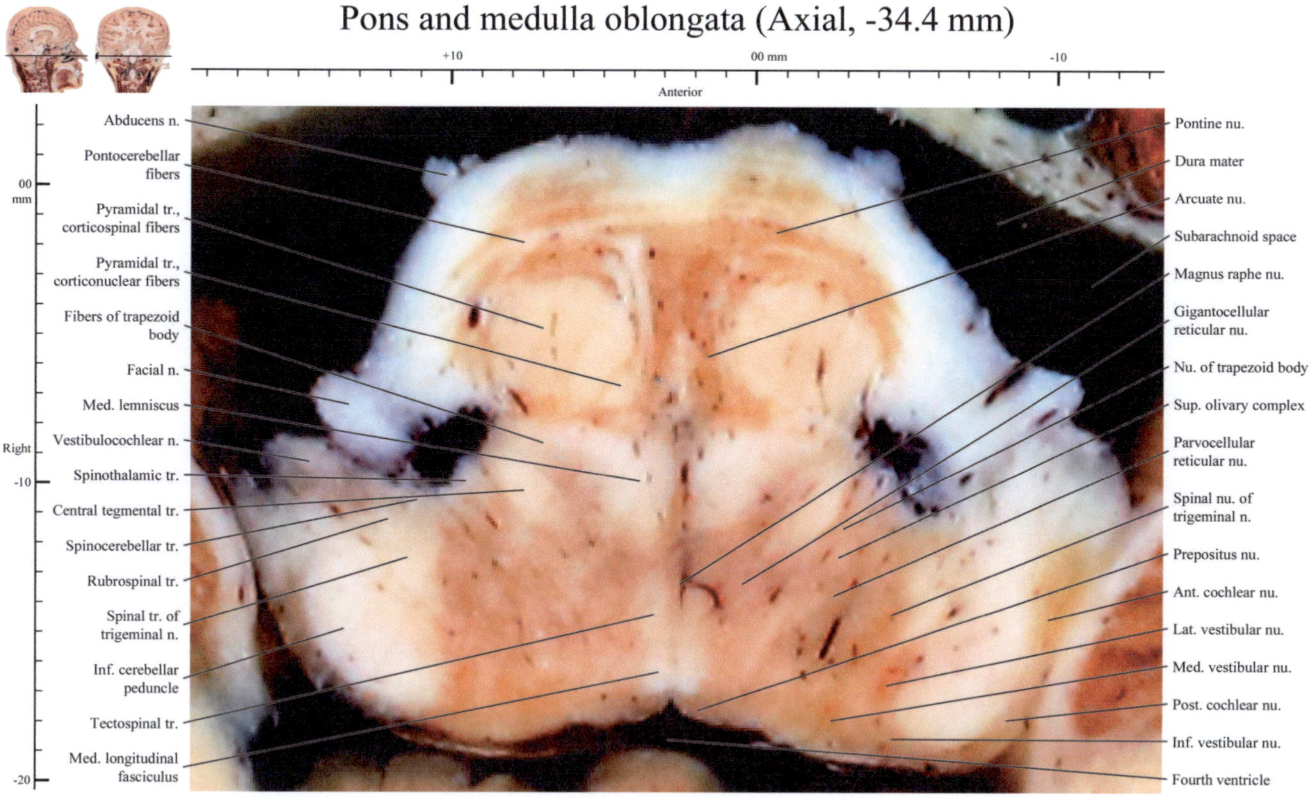

Pons and medulla oblongata (Axial, -35.0 mm)

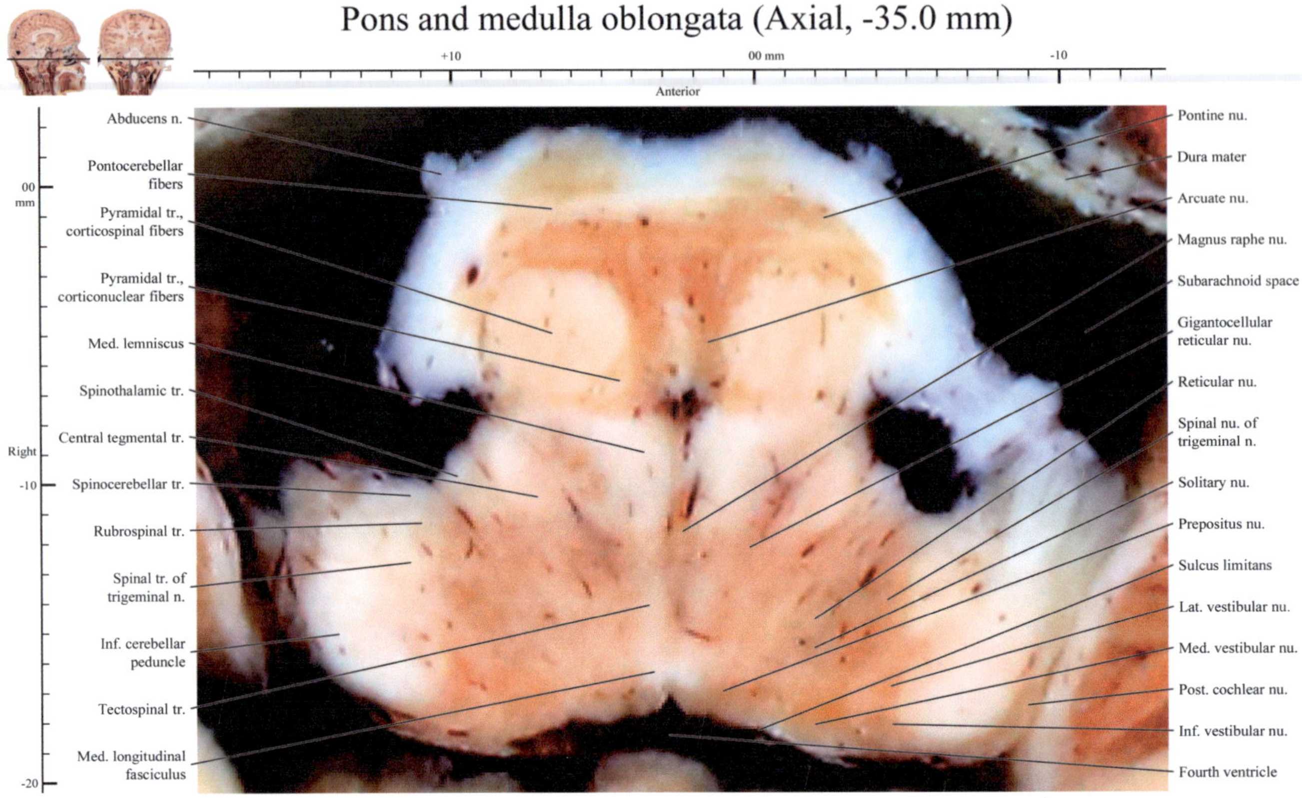

Pons and medulla oblongata (Axial, -35.4 mm)

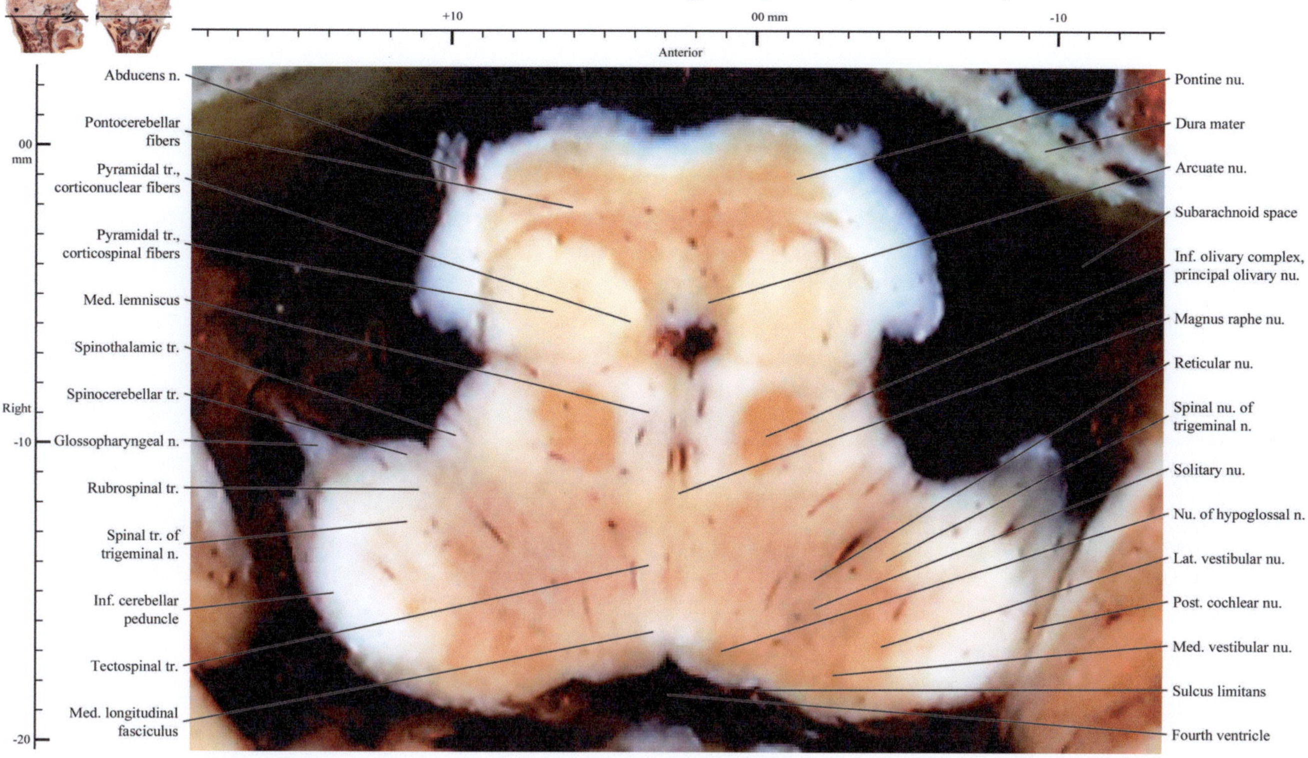

Pons and medulla oblongata (Axial, -36.0 mm)

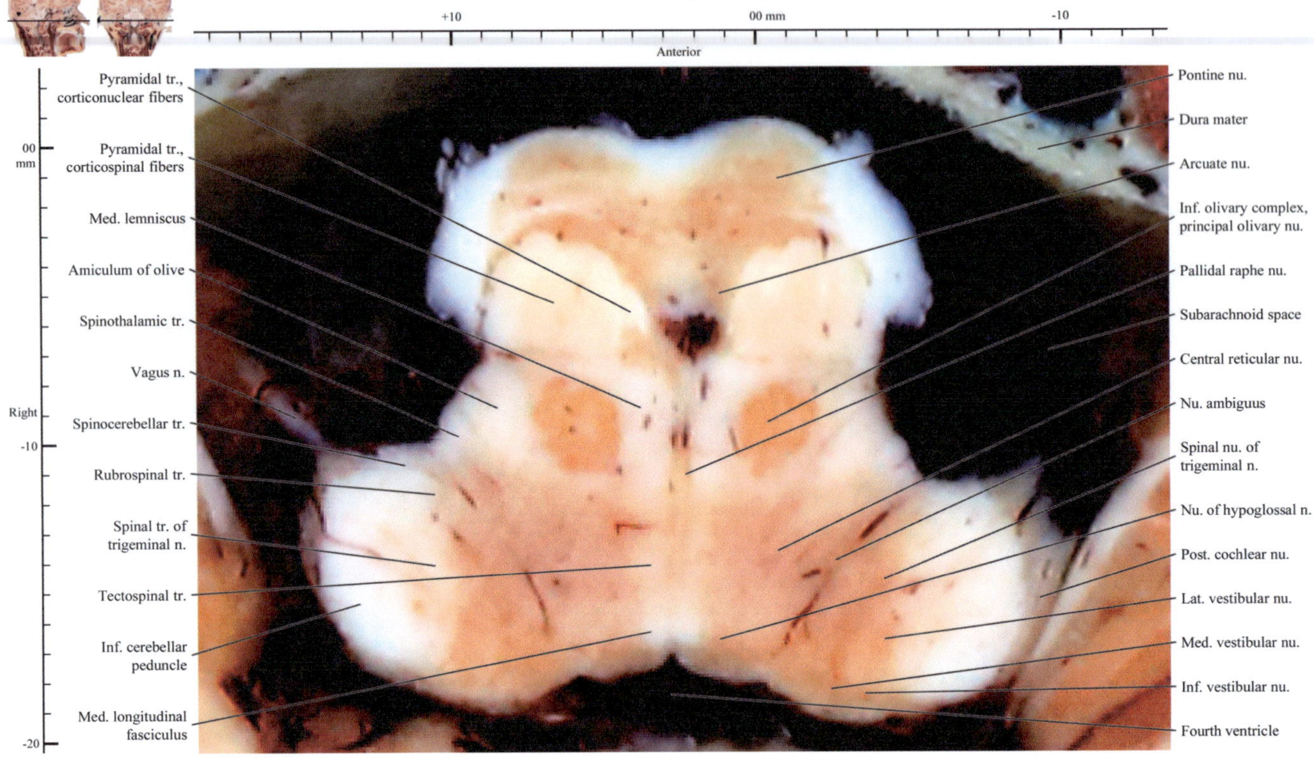

Pons and medulla oblongata (Axial, -36.4 mm)

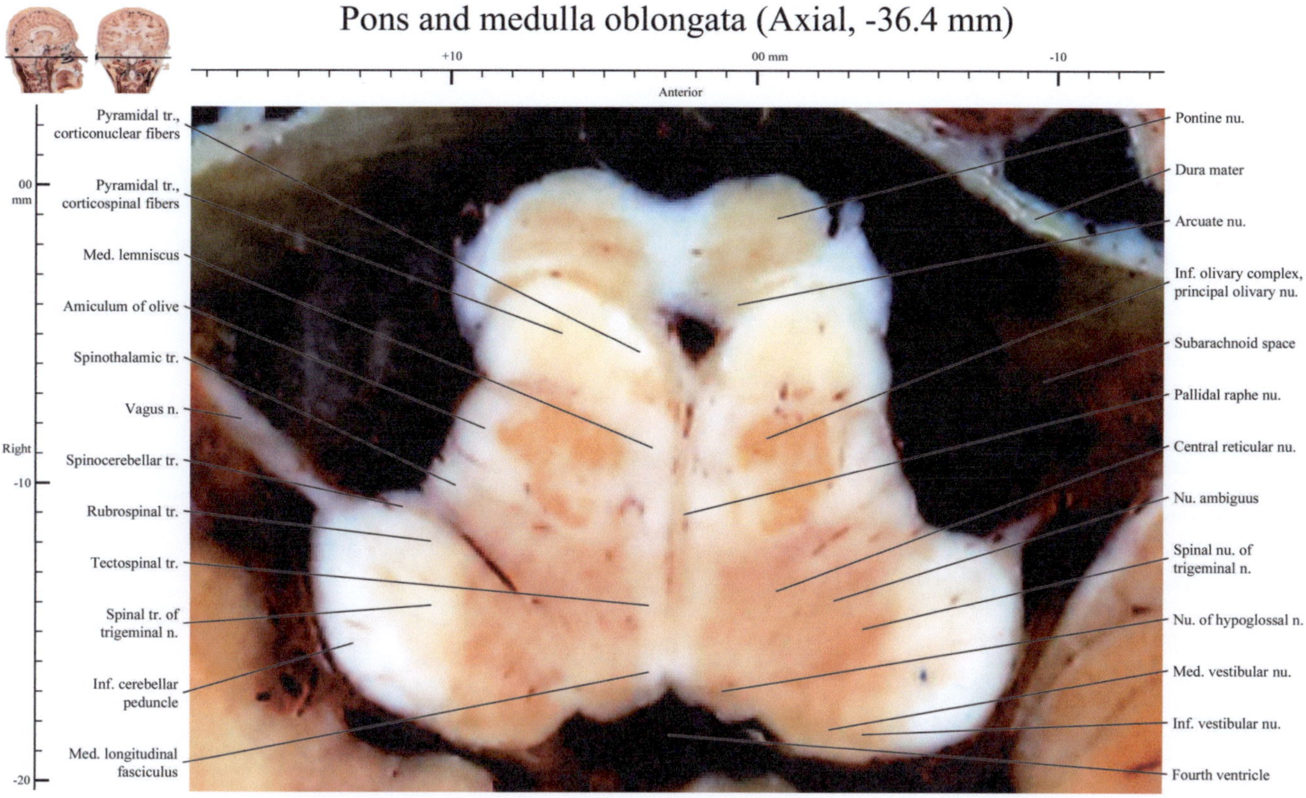

Medulla oblongata (Axial, -37.0 mm)

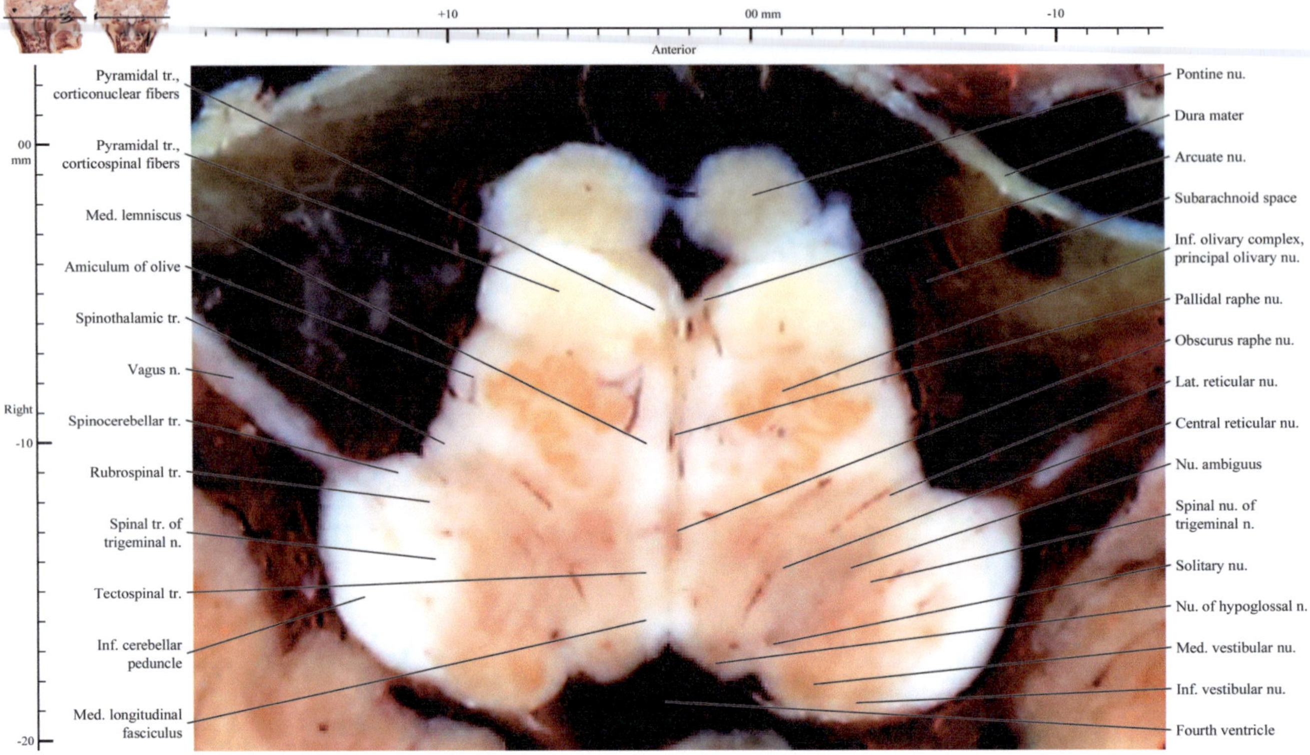

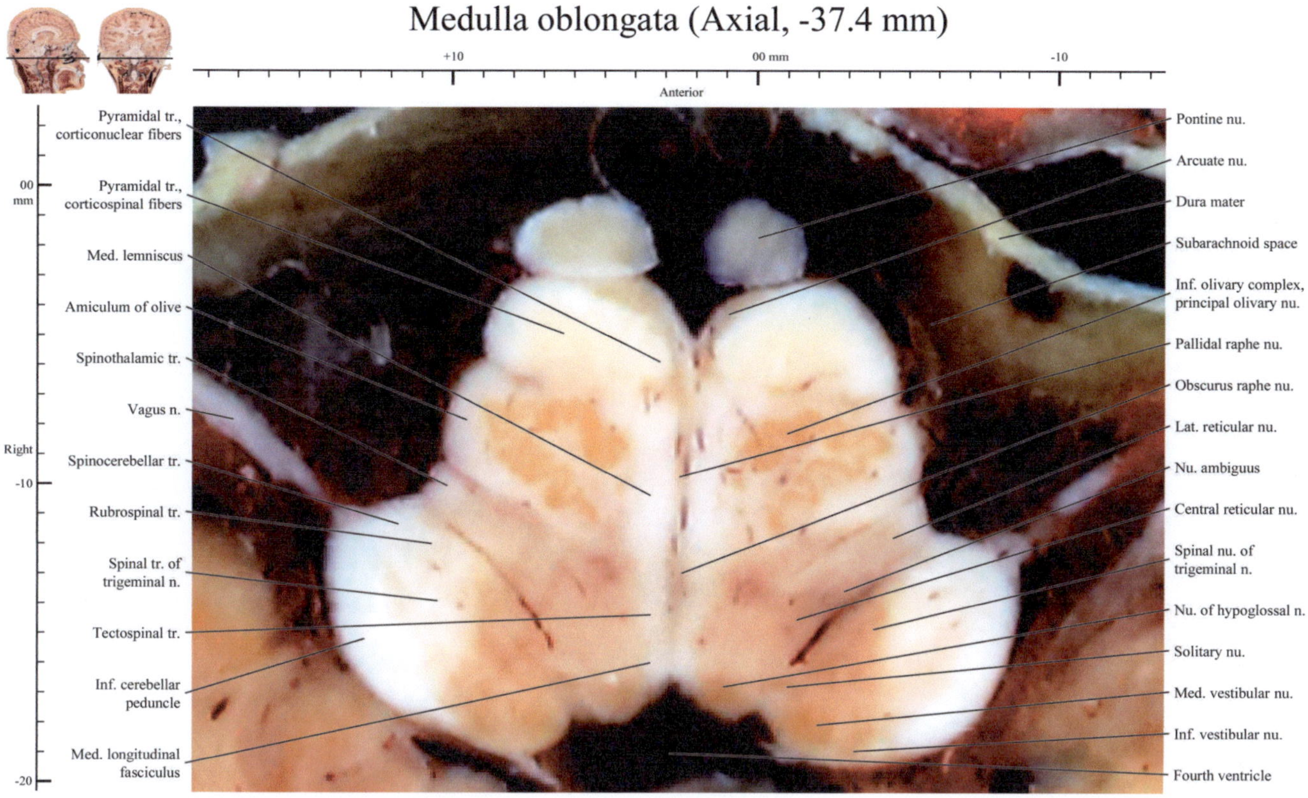
Medulla oblongata (Axial, -37.4 mm)
+10
00 mm
-10
Anterior
00 mm
Right
-10
-20
Pyramidal tr., corticonuclear fibers
Pyramidal tr., corticospinal fibers
Med. lemniscus
Amiculum of olive
Spinothalamic tr.
Vagus n.
Spinocerebellar tr.
Rubrospinal tr.
Spinal tr. of trigeminal n.
Tectospinal tr.
Inf. cerebellar peduncle
Med. longitudinal fasciculus
Pontine nu.
Arcuate nu.
Dura mater
Subarachnoid space
Inf. olivary complex, principal olivary nu.
Pallidal raphe nu.
Obscurus raphe nu.
Lat. reticular nu.
Nu. ambiguus
Central reticular nu.
Spinal nu. of trigeminal n.
Nu. of hypoglossal n.
Solitary nu.
Med. vestibular nu.
Inf. vestibular nu.
Fourth ventricle

Medulla oblongata (Axial, -38.0 mm)

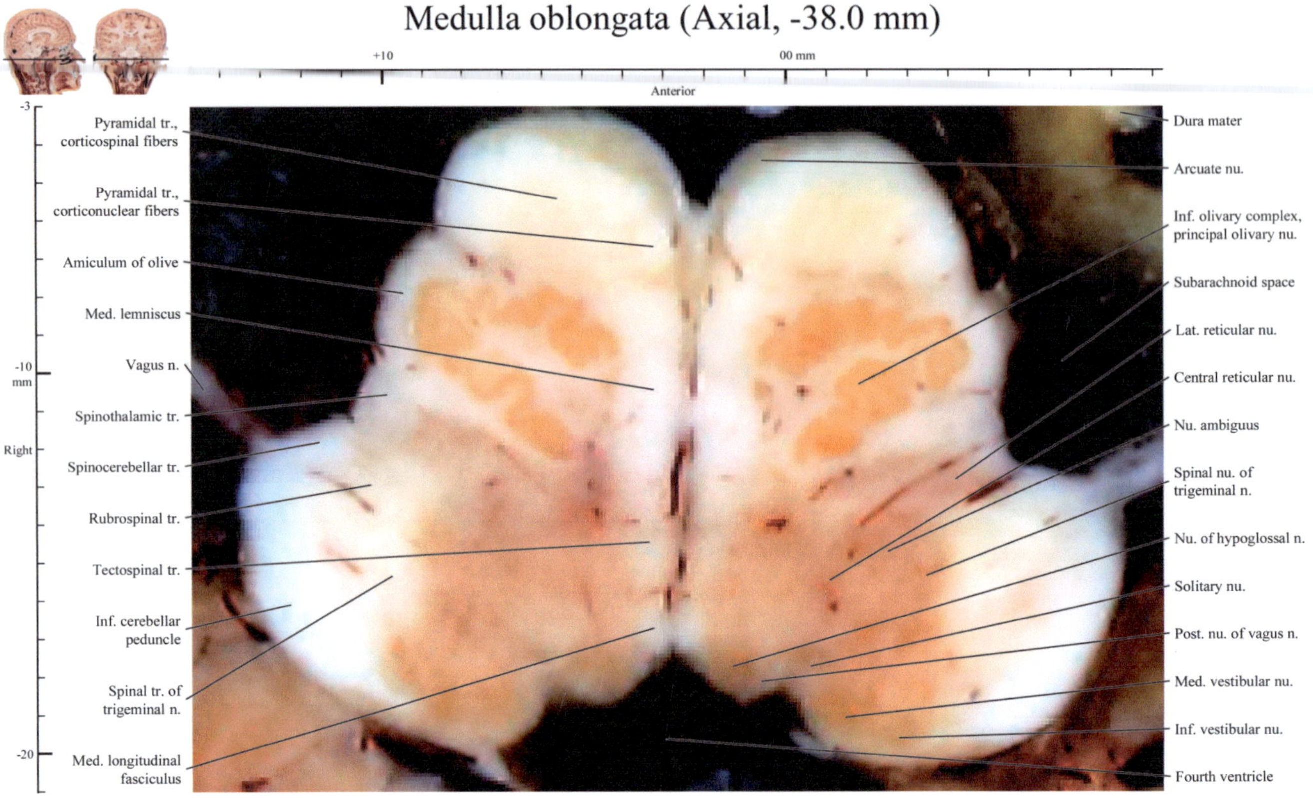

Medulla oblongata (Axial, -38.4 mm)

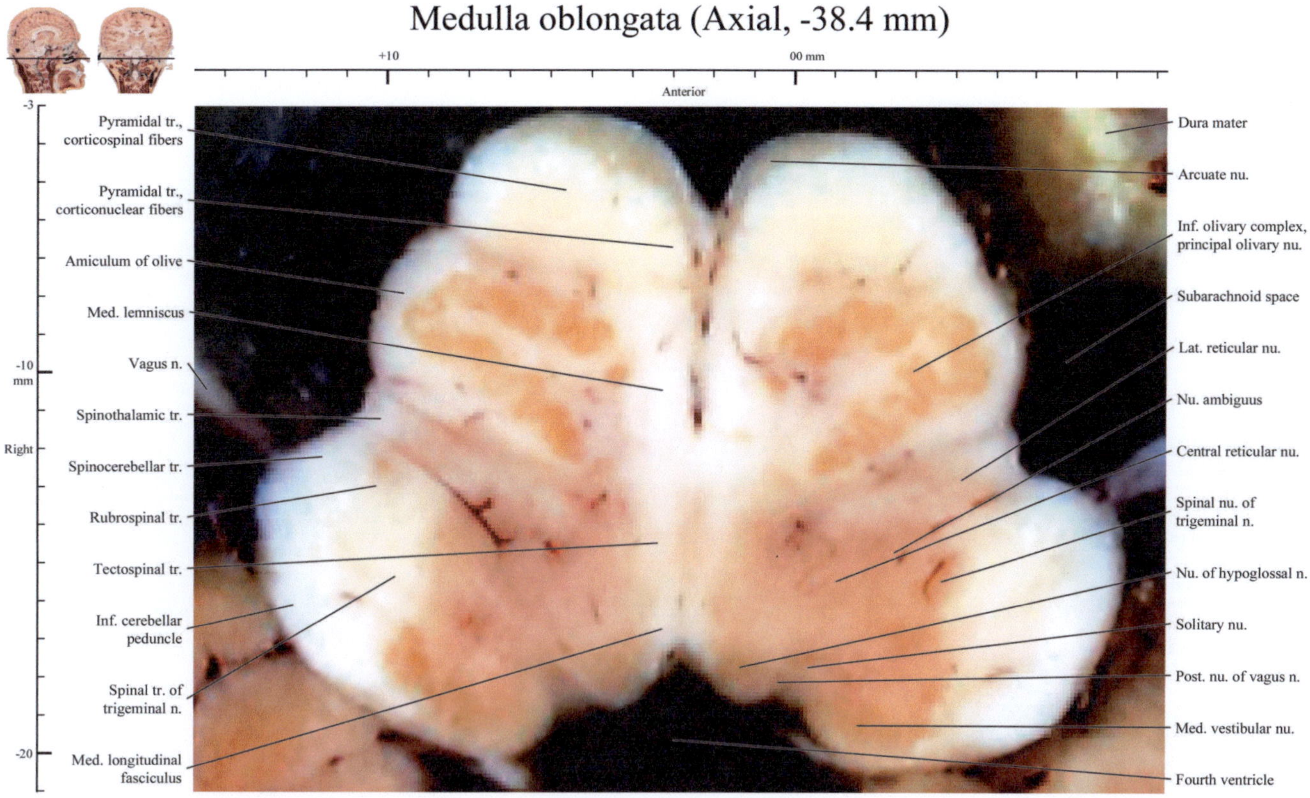

Medulla oblongata (Axial, -39.0 mm)

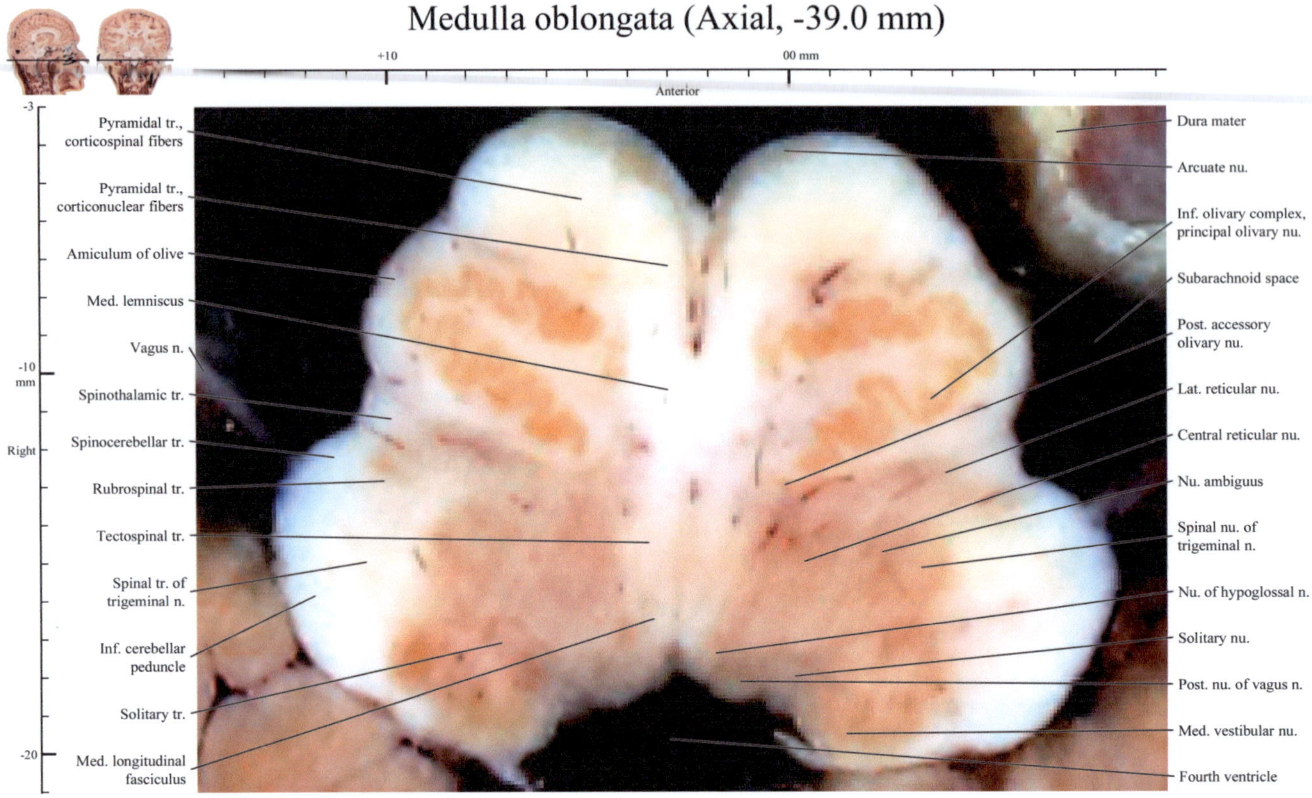

Medulla oblongata (Axial, -39.4 mm)

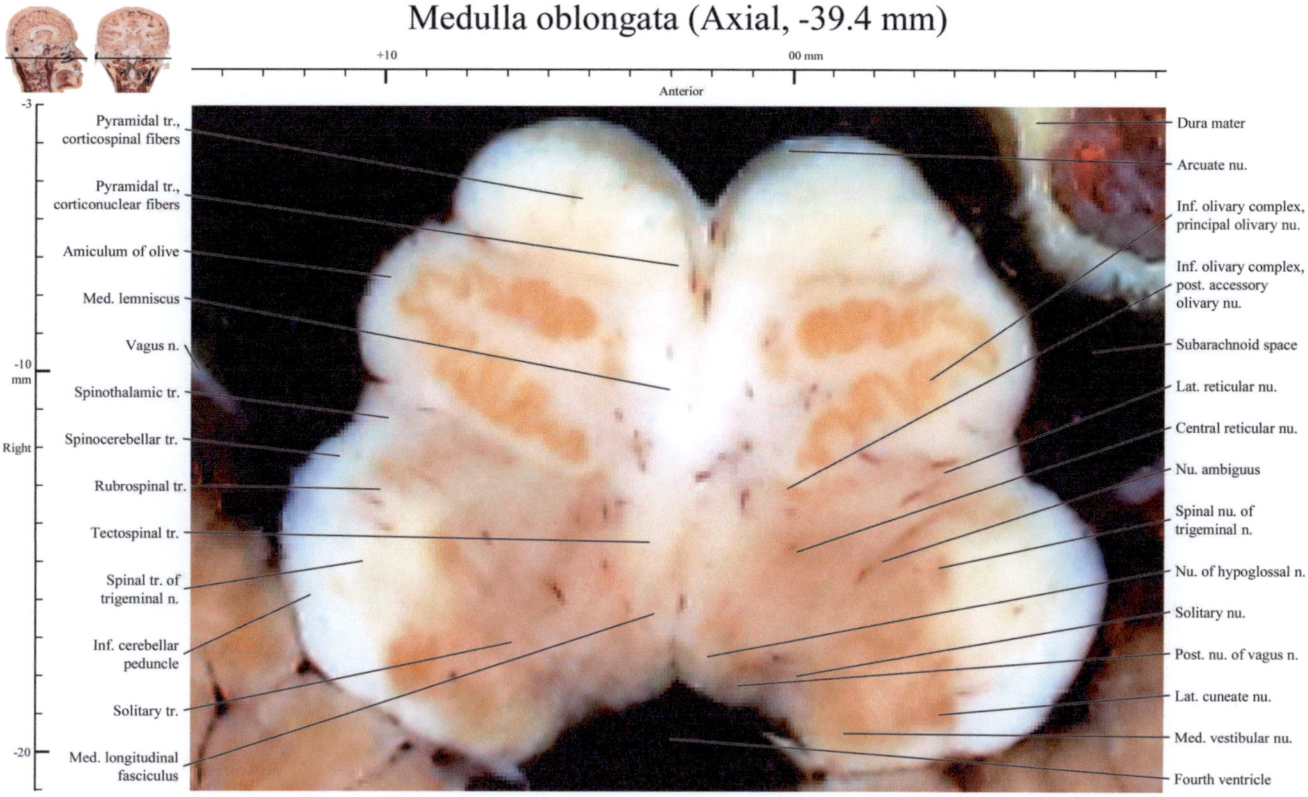

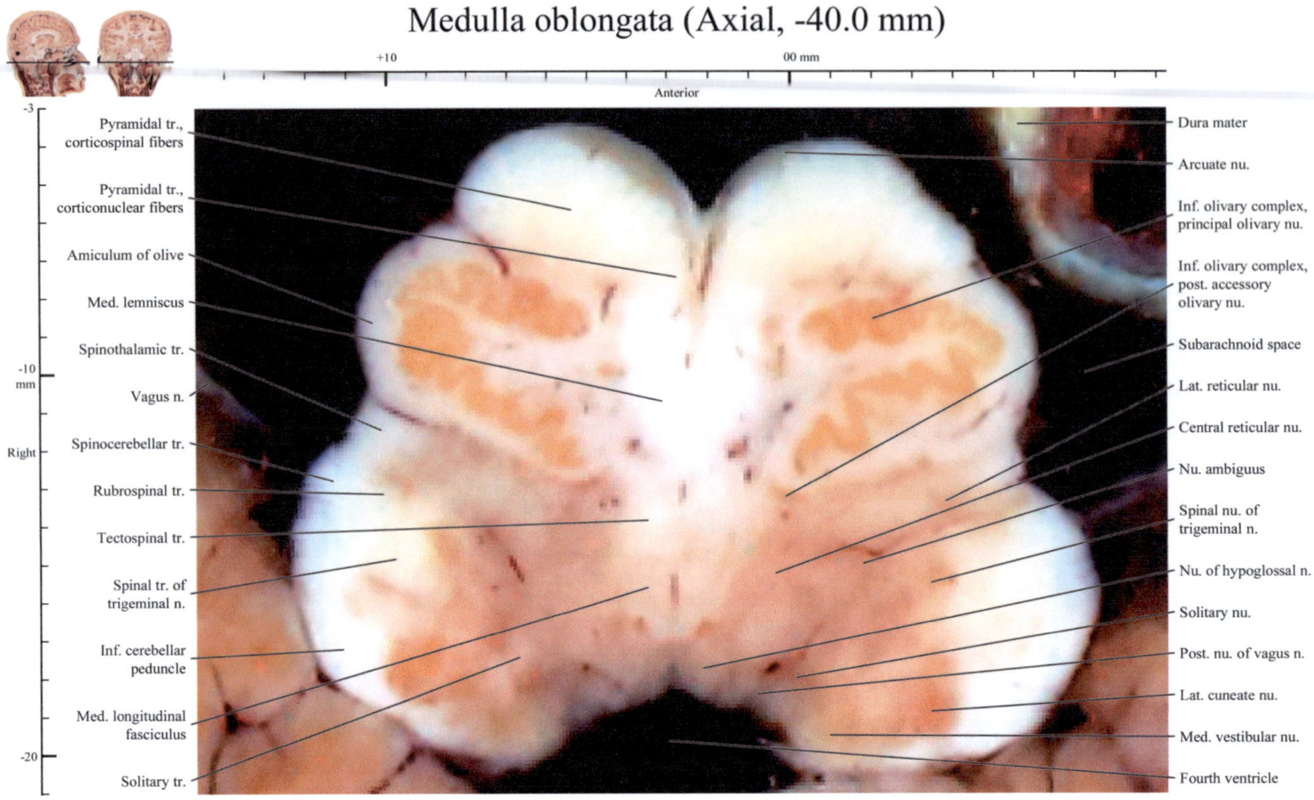
Medulla oblongata (Axial, -40.0 mm)
+10
00 mm
Anterior
-3
-10 mm
Right
-20
Pyramidal tr., corticospinal fibers
Pyramidal tr., corticonuclear fibers
Amiculum of olive
Med. lemniscus
Spinothalamic tr.
Vagus n.
Spinocerebellar tr.
Rubrospinal tr.
Tectospinal tr.
Spinal tr. of trigeminal n.
Inf. cerebellar peduncle
Med. longitudinal fasciculus
Solitary tr.
Dura mater
Arcuate nu.
Inf. olivary complex, principal olivary nu.
Inf. olivary complex, post. accessory olivary nu.
Subarachnoid space
Lat. reticular nu.
Central reticular nu.
Nu. ambiguus
Spinal nu. of trigeminal n.
Nu. of hypoglossal n.
Solitary nu.
Post. nu. of vagus n.
Lat. cuneate nu.
Med. vestibular nu.
Fourth ventricle

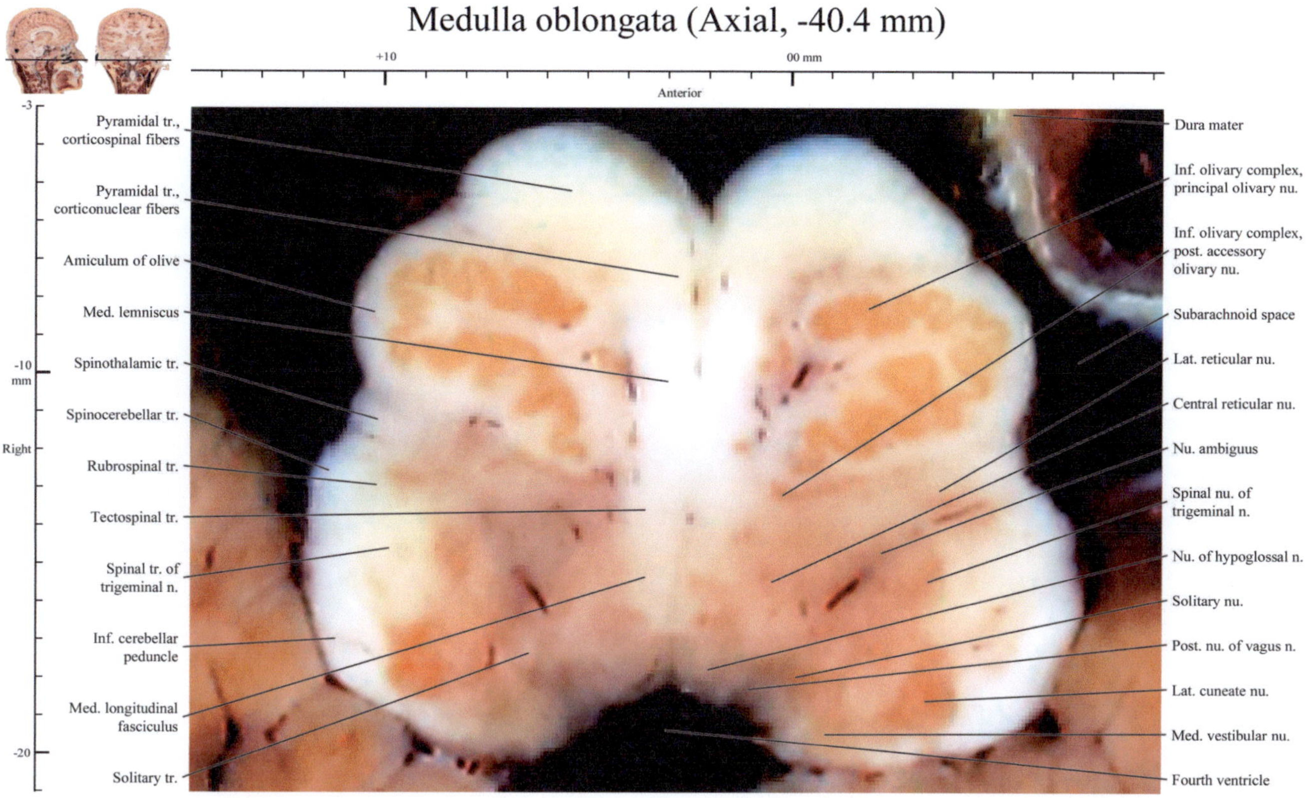
Medulla oblongata (Axial, -40.4 mm)
+10
00 mm
Anterior
-3
-10 mm
Right
-20
Pyramidal tr., corticospinal fibers
Pyramidal tr., corticonuclear fibers
Amiculum of olive
Med. lemniscus
Spinothalamic tr.
Spinocerebellar tr.
Rubrospinal tr.
Tectospinal tr.
Spinal tr. of trigeminal n.
Inf. cerebellar peduncle
Med. longitudinal fasciculus
Solitary tr.
Dura mater
Inf. olivary complex, principal olivary nu.
Inf. olivary complex, post. accessory olivary nu.
Subarachnoid space
Lat. reticular nu.
Central reticular nu.
Nu. ambiguus
Spinal nu. of trigeminal n.
Nu. of hypoglossal n.
Solitary nu.
Post. nu. of vagus n.
Lat. cuneate nu.
Med. vestibular nu.
Fourth ventricle

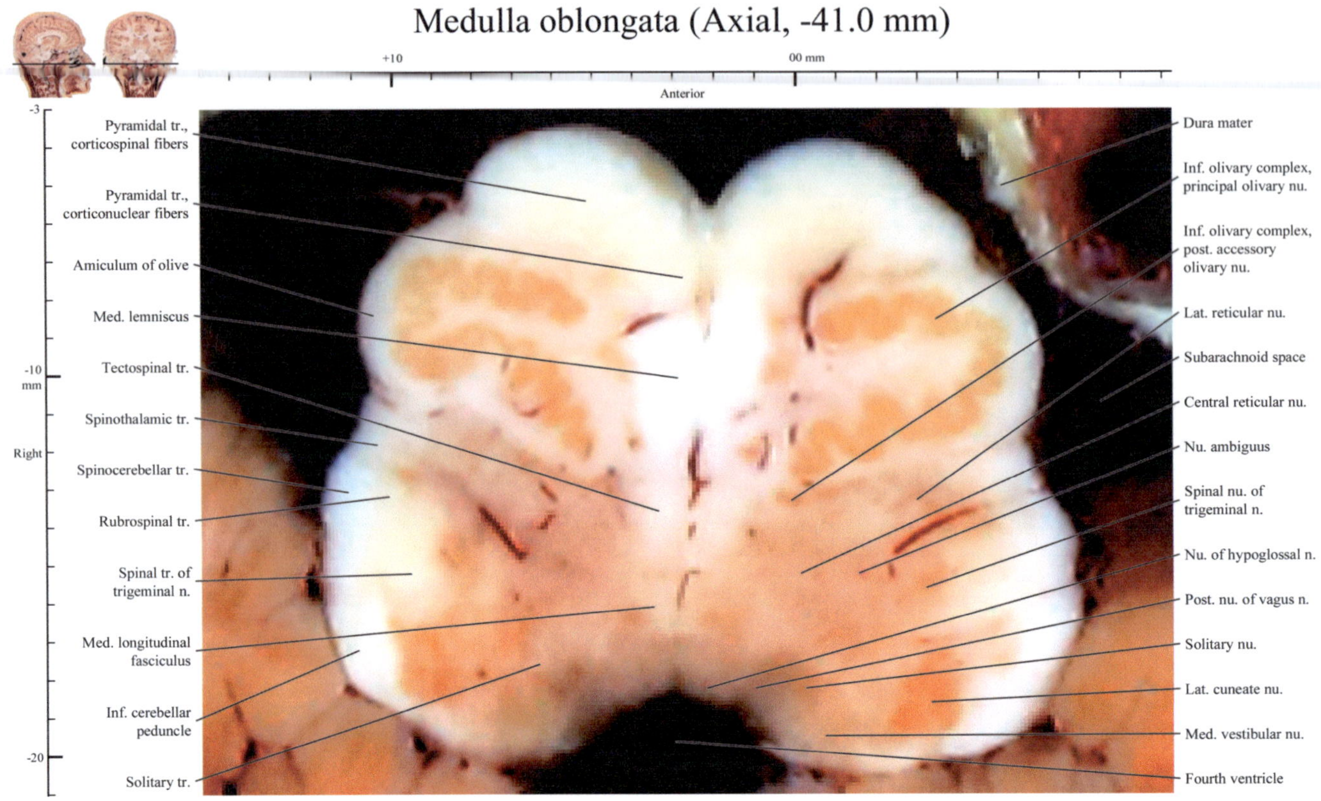

Medulla oblongata (Axial, -41.0 mm)
+10
00 mm
Anterior
-3
-10
mm
Right
-20
Pyramidal tr., corticospinal fibers
Pyramidal tr., corticonuclear fibers
Amiculum of olive
Med. lemniscus
Tectospinal tr.
Spinothalamic tr.
Spinocerebellar tr.
Rubrospinal tr.
Spinal tr. of trigeminal n.
Med. longitudinal fasciculus
Inf. cerebellar peduncle
Solitary tr.
Dura mater
Inf. olivary complex, principal olivary nu.
Inf. olivary complex, post. accessory olivary nu.
Lat. reticular nu.
Subarachnoid space
Central reticular nu.
Nu. ambiguus
Spinal nu. of trigeminal n.
Nu. of hypoglossal n.
Post. nu. of vagus n.
Solitary nu.
Lat. cuneate nu.
Med. vestibular nu.
Fourth ventricle

Medulla oblongata (Axial, -41.4 mm)

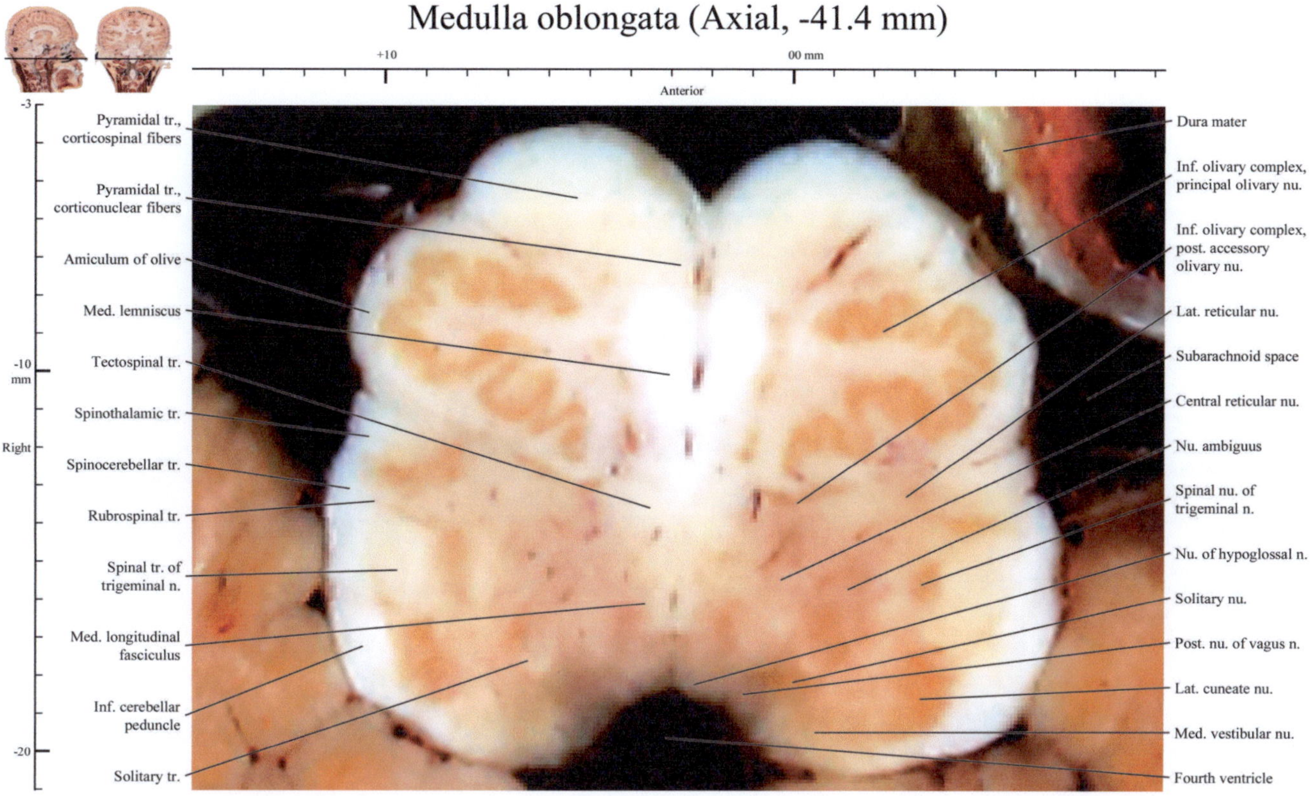

Medulla oblongata (Axial, -42.0 mm)

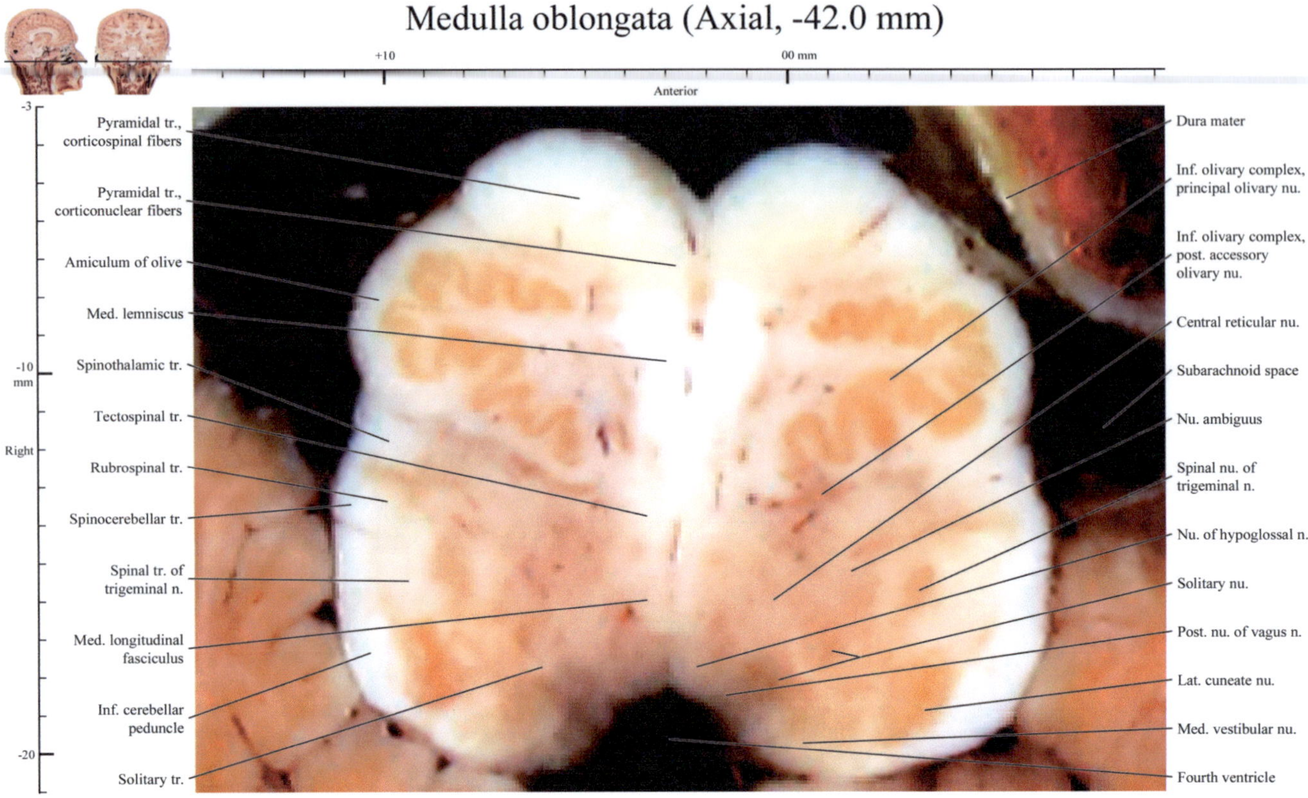

Medulla oblongata (Axial, -42.4 mm)

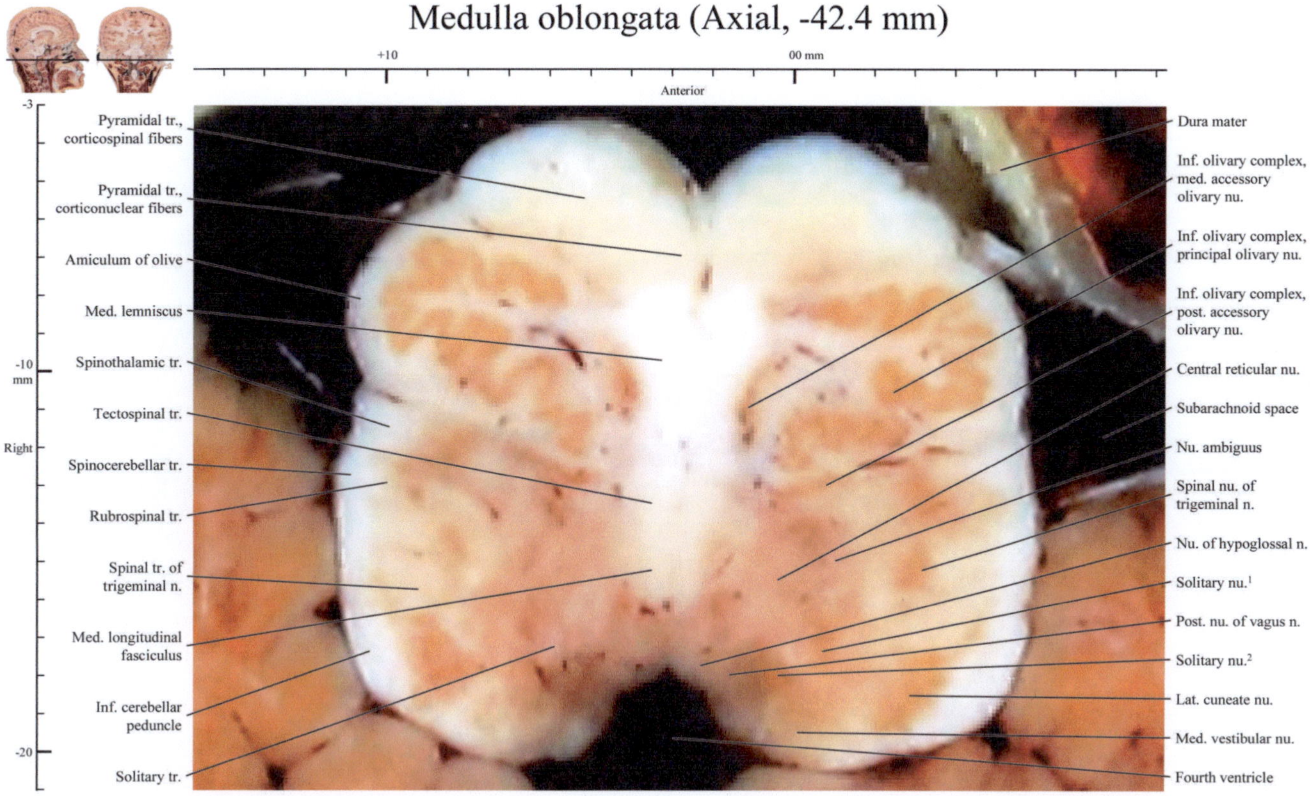

Medulla oblongata (Axial, -43.0 mm)

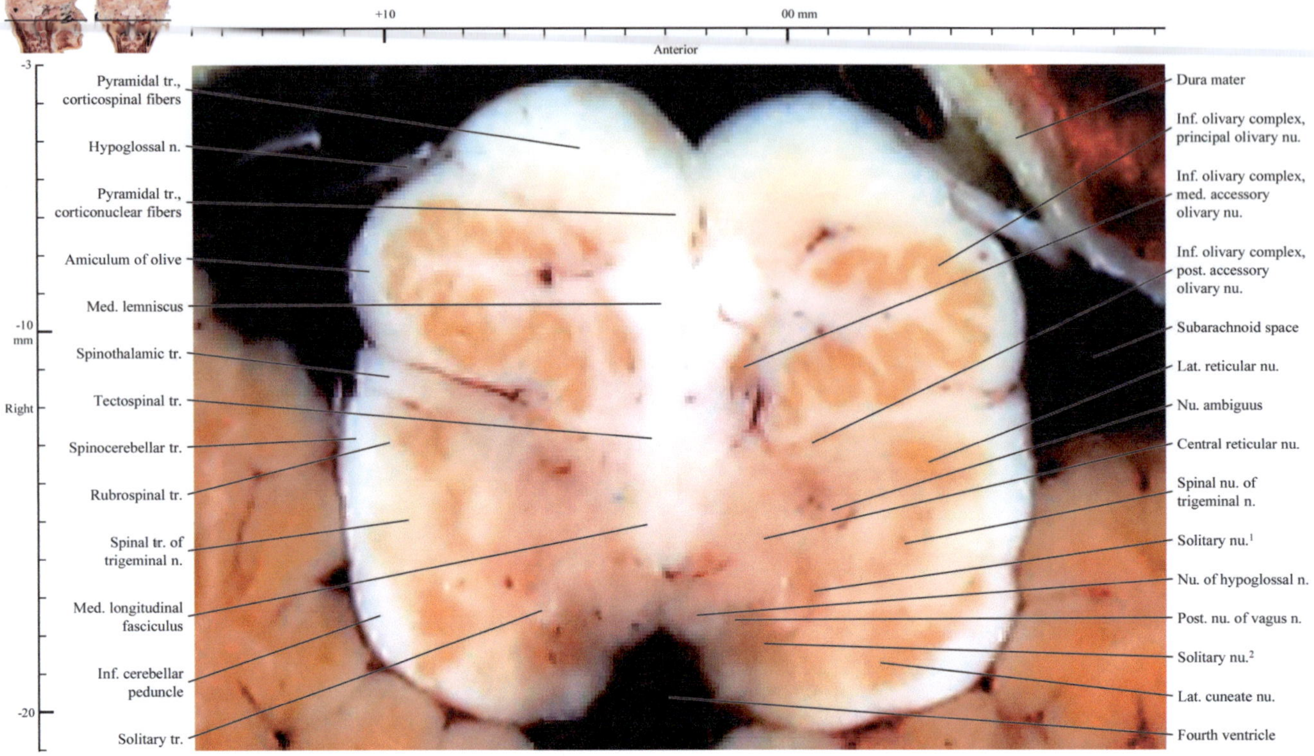

Medulla oblongata (Axial, -43.4 mm)

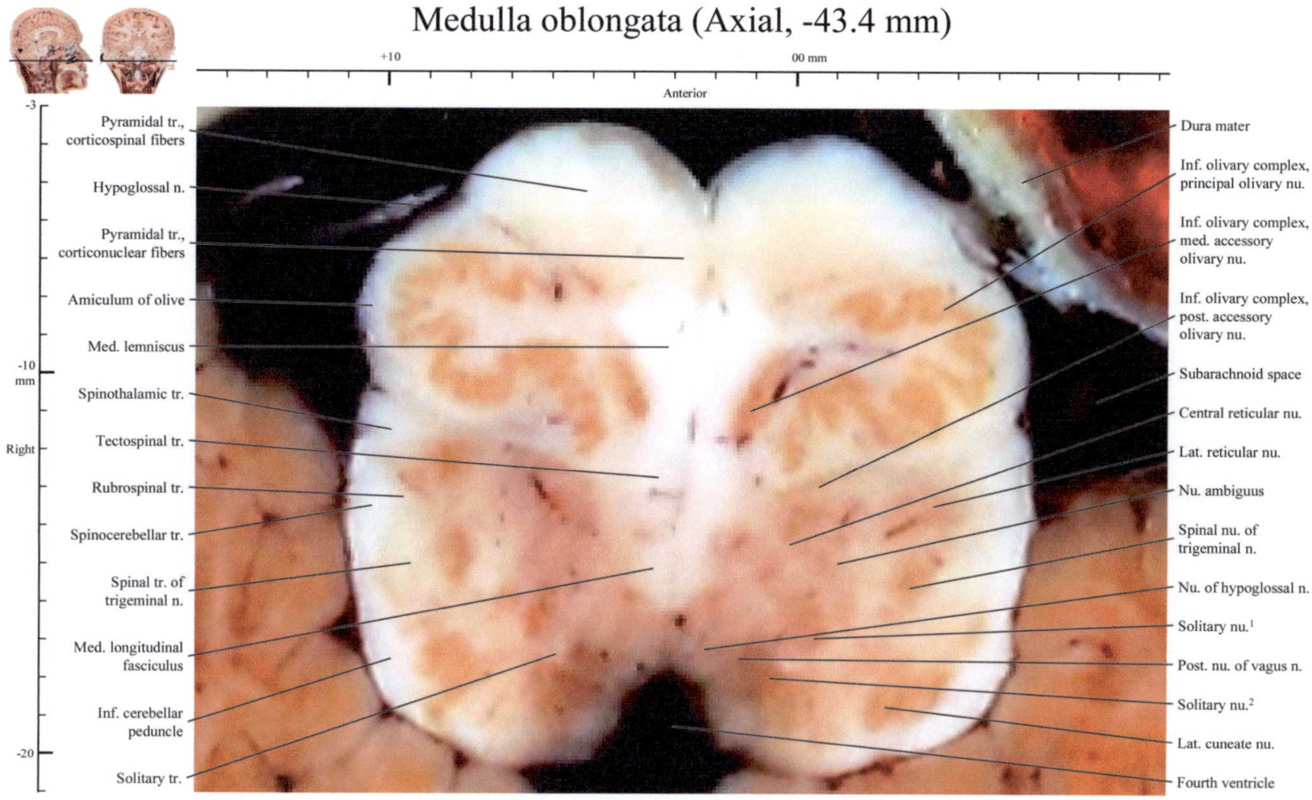

Medulla oblongata (Axial, -44.0 mm)

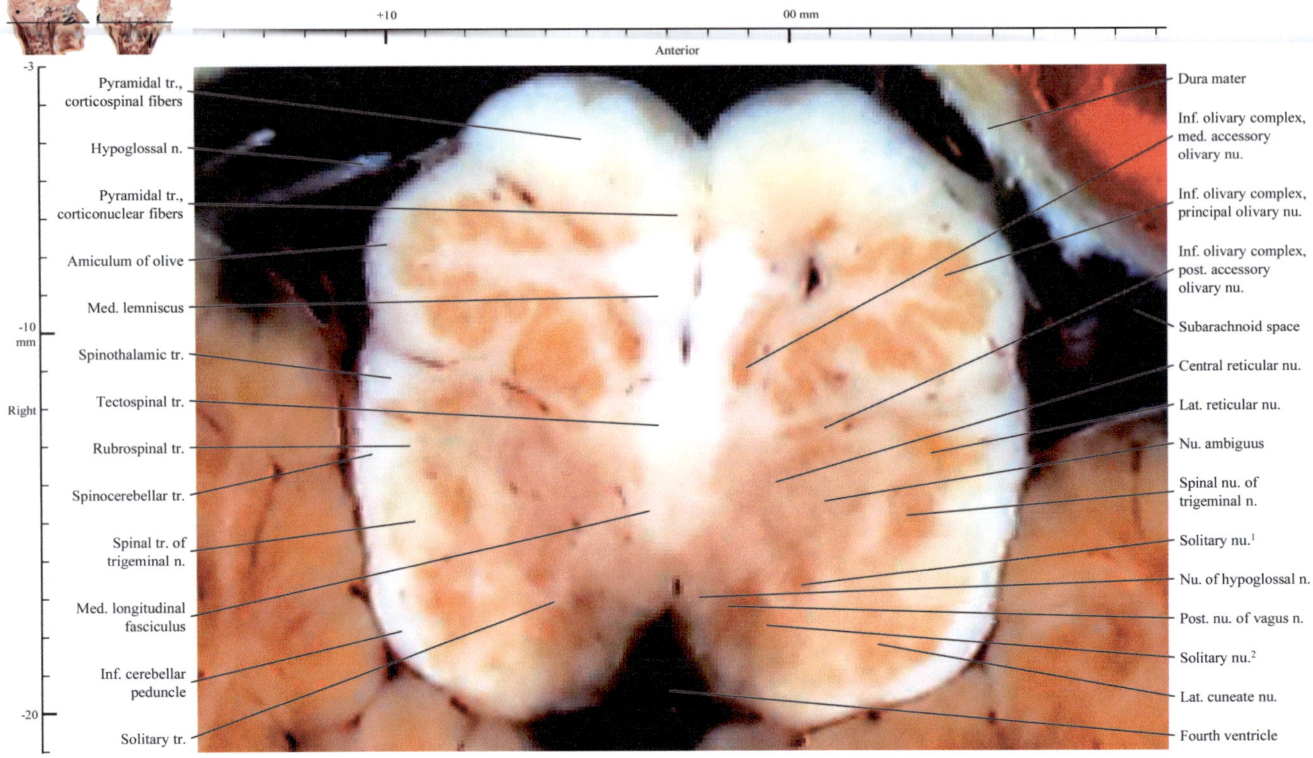

Medulla oblongata (Axial, -44.4 mm)

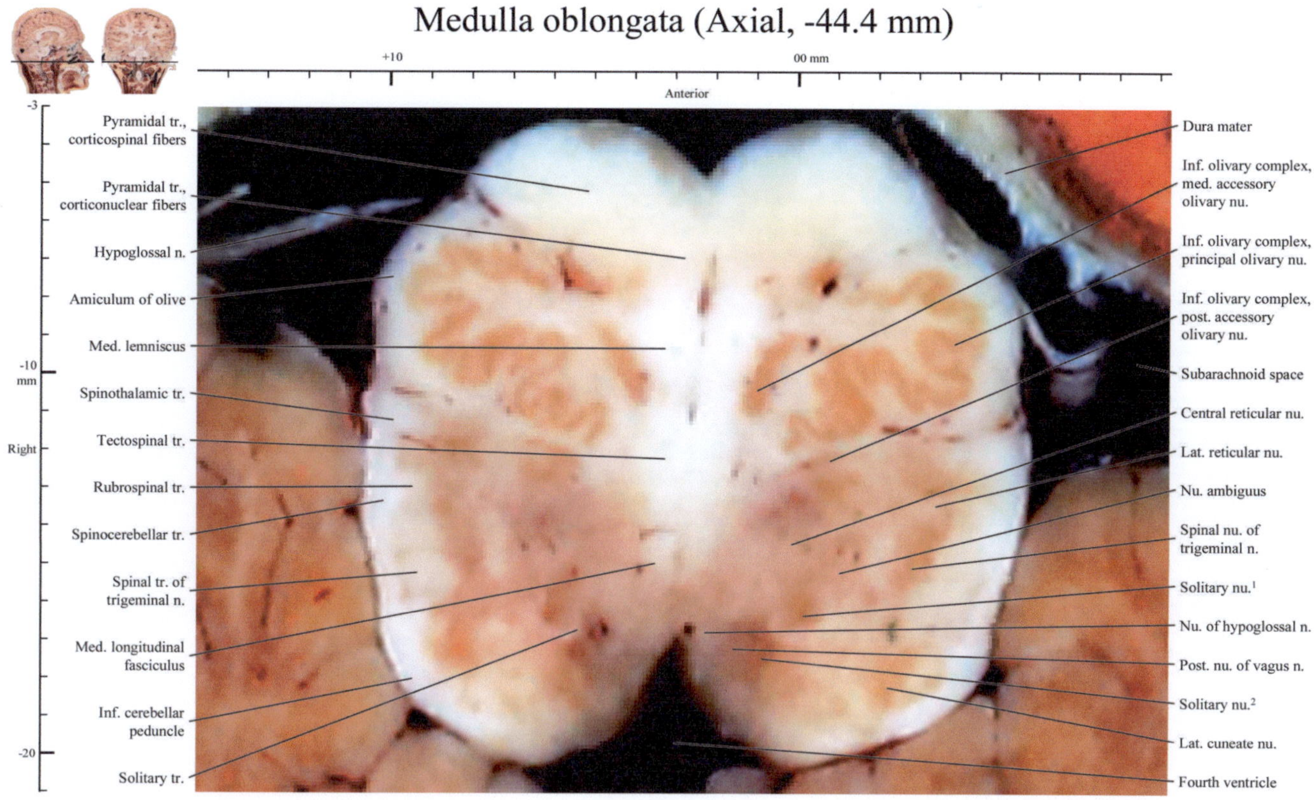

Medulla oblongata (Axial, -45.0 mm)

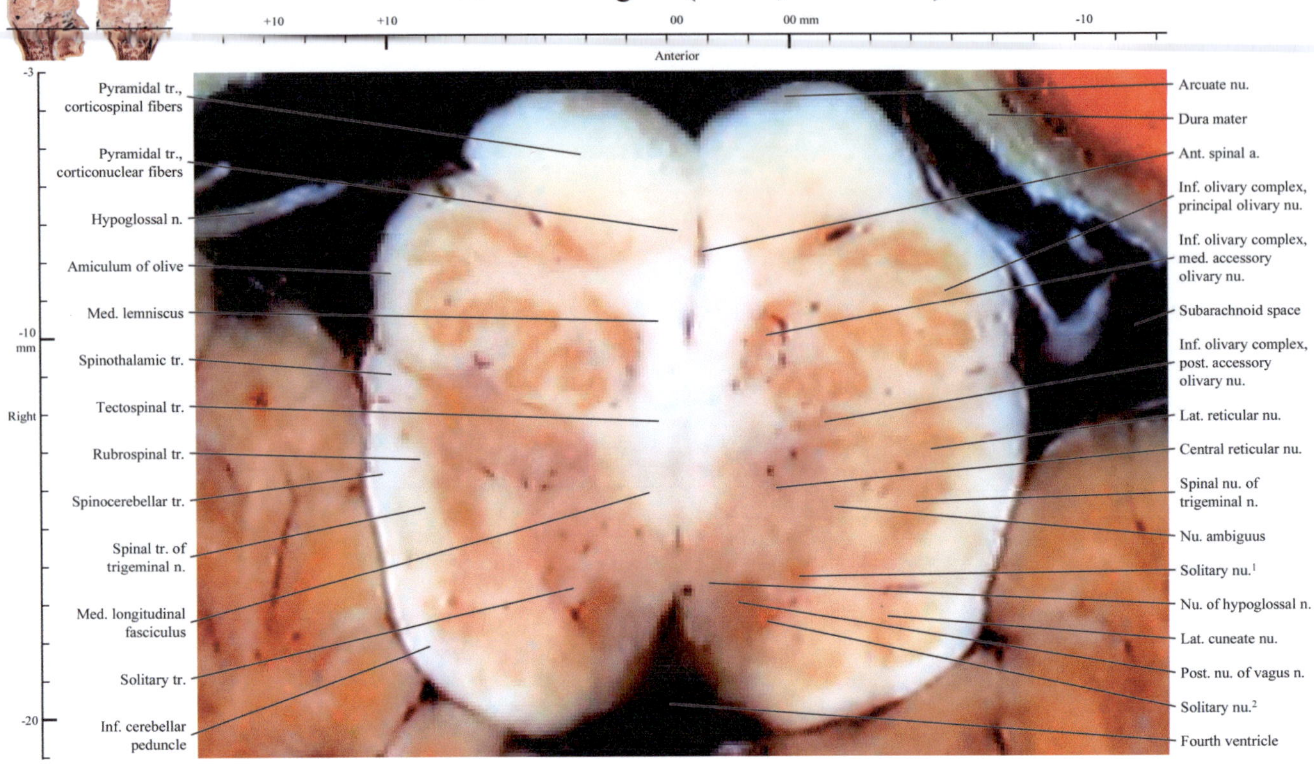

Medulla oblongata (Axial, -45.4 mm)

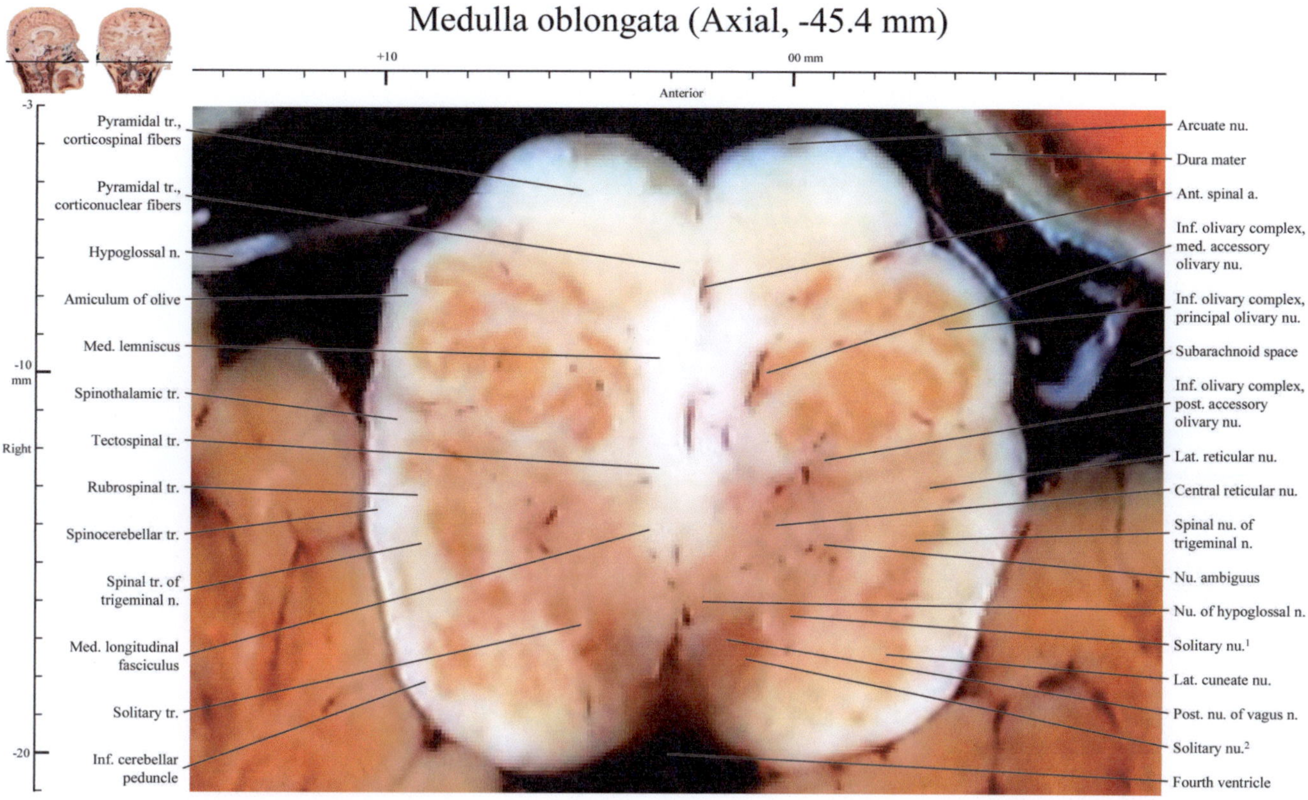

Medulla oblongata (Axial, -46.0 mm)

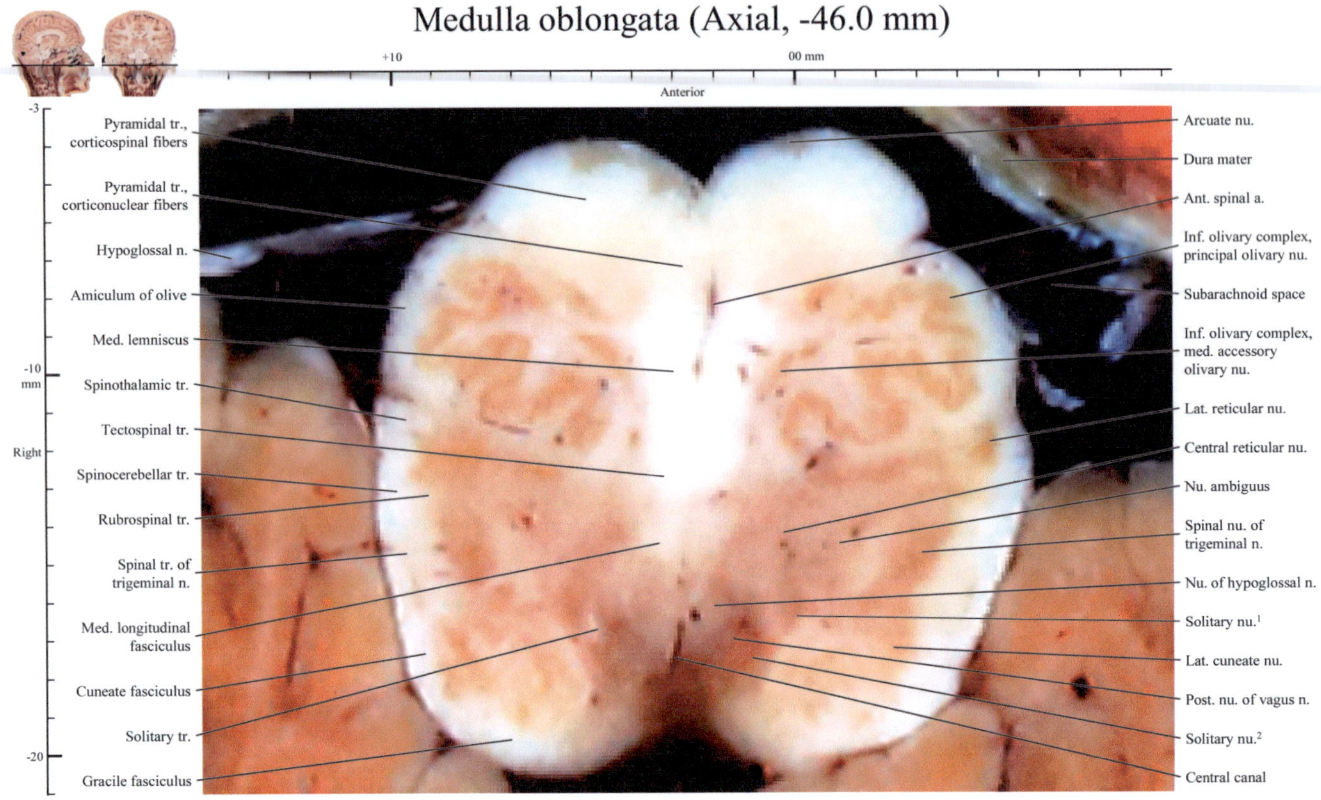

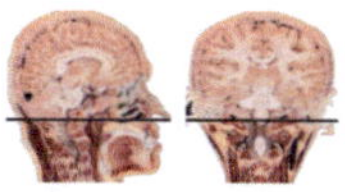

Medulla oblongata (Axial, -46.4 mm)

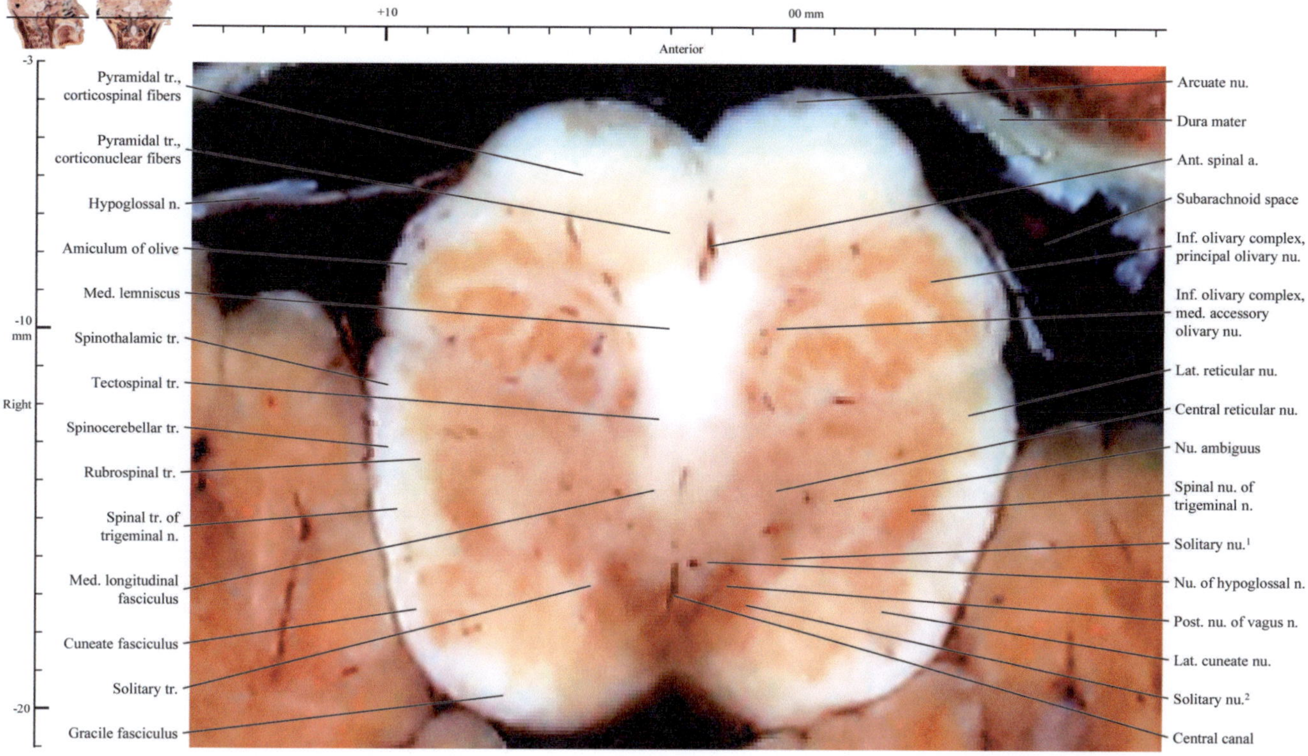

Medulla oblongata (Axial, -47.0 mm)

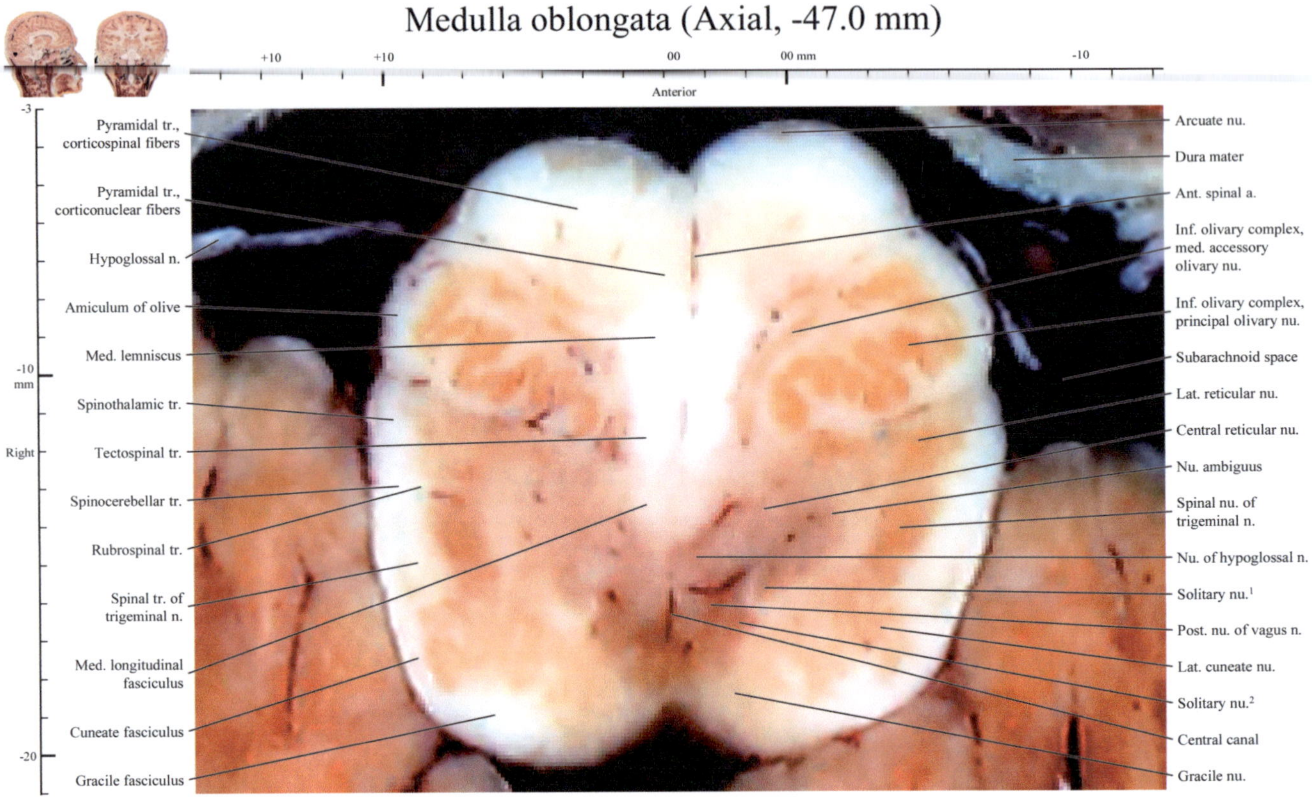

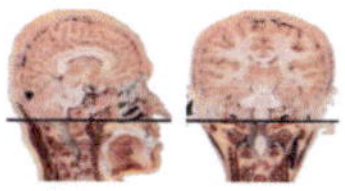

Medulla oblongata (Axial, -47.4 mm)

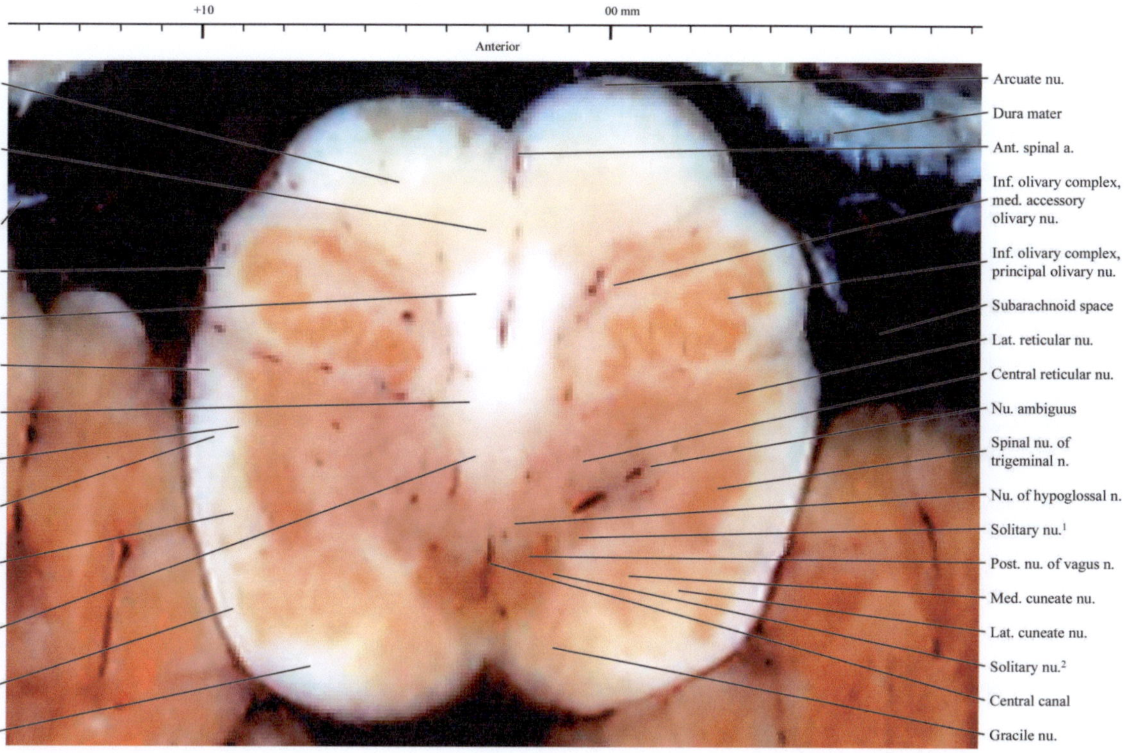

Medulla oblongata (Axial, -48.0 mm)

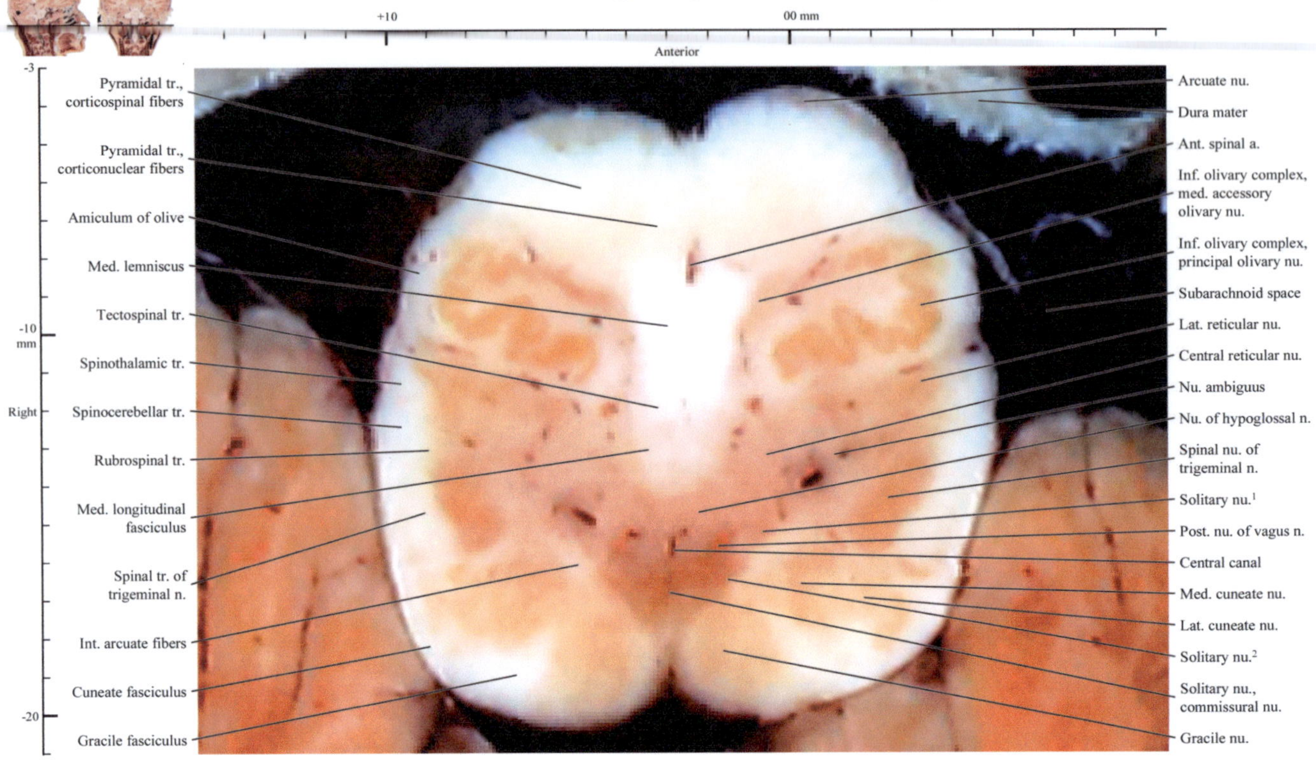

Medulla oblongata (Axial, -48.4 mm)

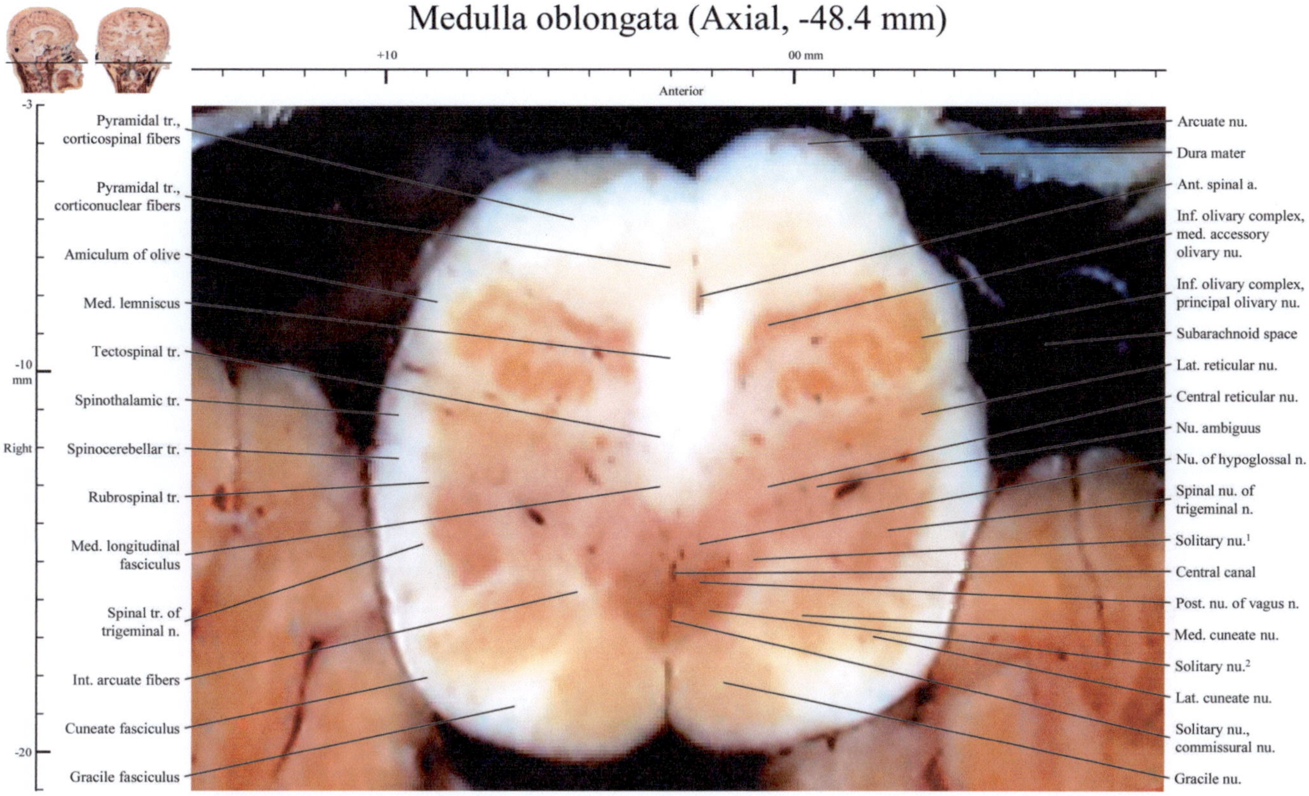

Medulla oblongata (Axial, -49.0 mm)

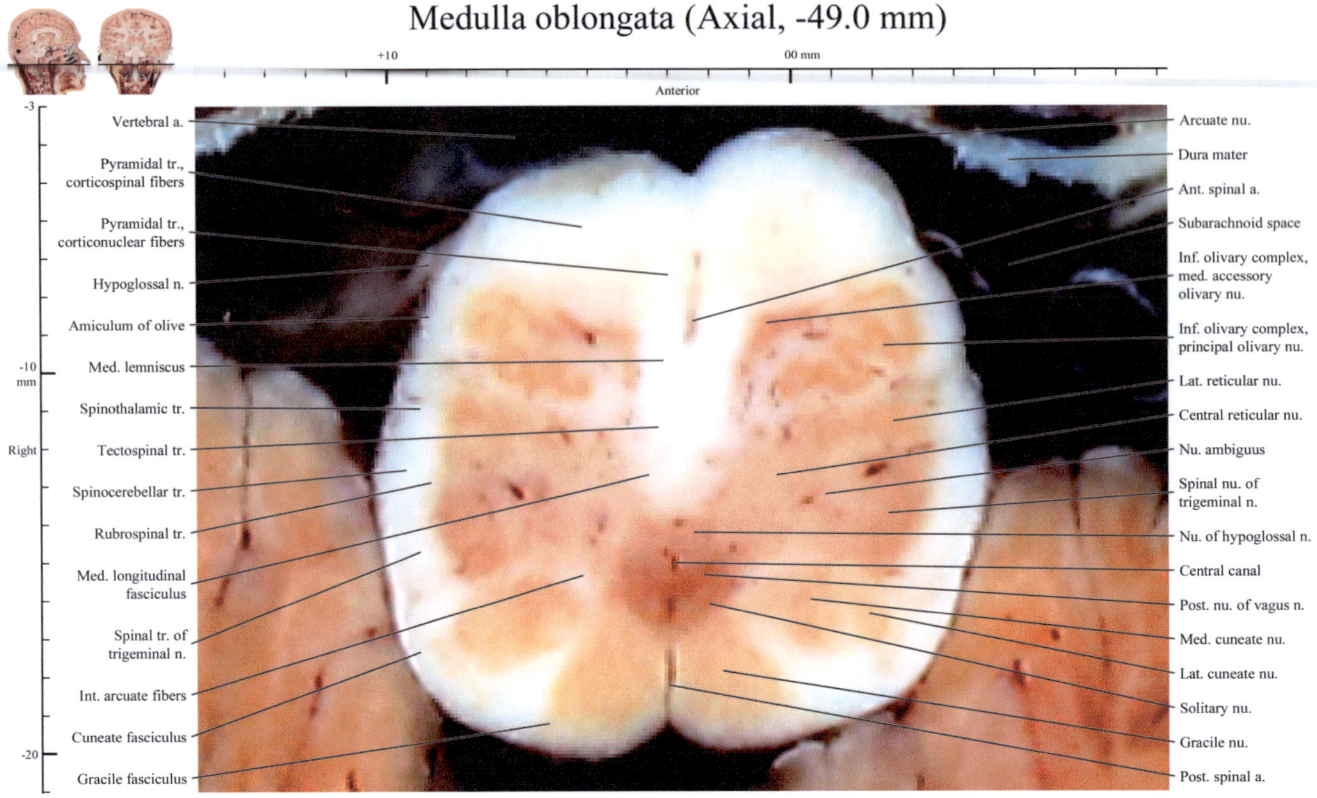

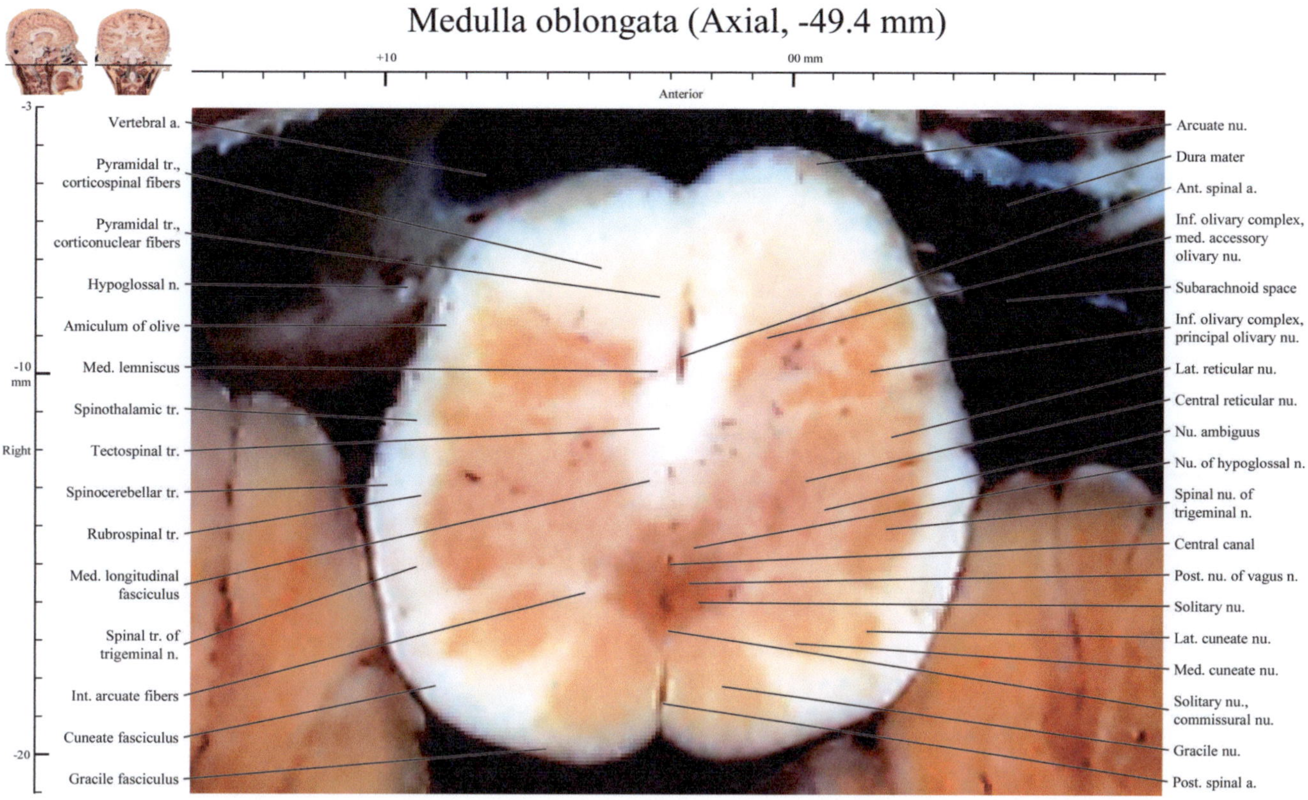

Medulla oblongata (Axial, -49.4 mm)
+10
00 mm
Anterior
-3
-10 mm
Right
-20
Vertebral a.
Pyramidal tr., corticospinal fibers
Pyramidal tr., corticonuclear fibers
Hypoglossal n.
Amiculum of olive
Med. lemniscus
Spinothalamic tr.
Tectospinal tr.
Spinocerebellar tr.
Rubrospinal tr.
Med. longitudinal fasciculus
Spinal tr. of trigeminal n.
Int. arcuate fibers
Cuneate fasciculus
Gracile fasciculus
Arcuate nu.
Dura mater
Ant. spinal a.
Inf. olivary complex, med. accessory olivary nu.
Subarachnoid space
Inf. olivary complex, principal olivary nu.
Lat. reticular nu.
Central reticular nu.
Nu. ambiguus
Nu. of hypoglossal n.
Spinal nu. of trigeminal n.
Central canal
Post. nu. of vagus n.
Solitary nu.
Lat. cuneate nu.
Med. cuneate nu.
Solitary nu., commissural nu.
Gracile nu.
Post. spinal a.

Medulla oblongata (Axial, -50.0 mm)

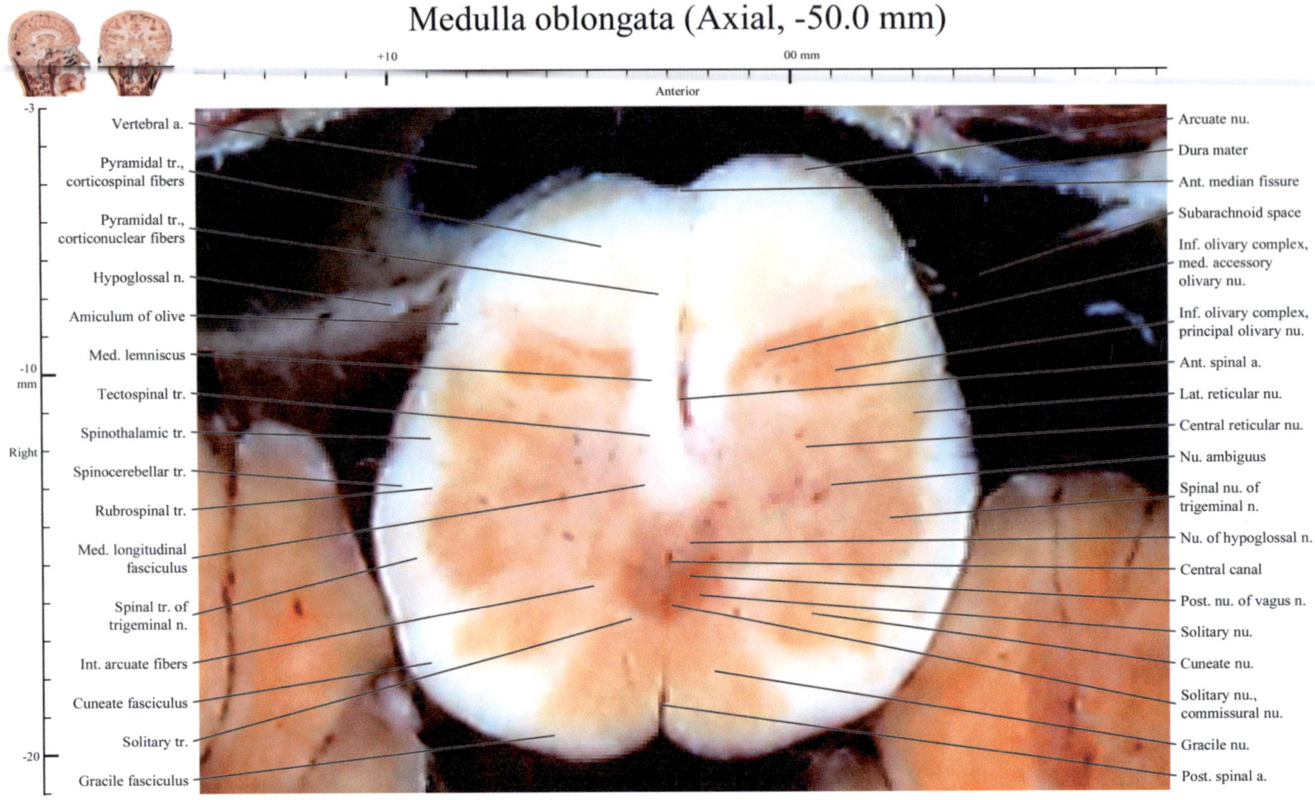

Medulla oblongata (Axial, -50.4 mm)

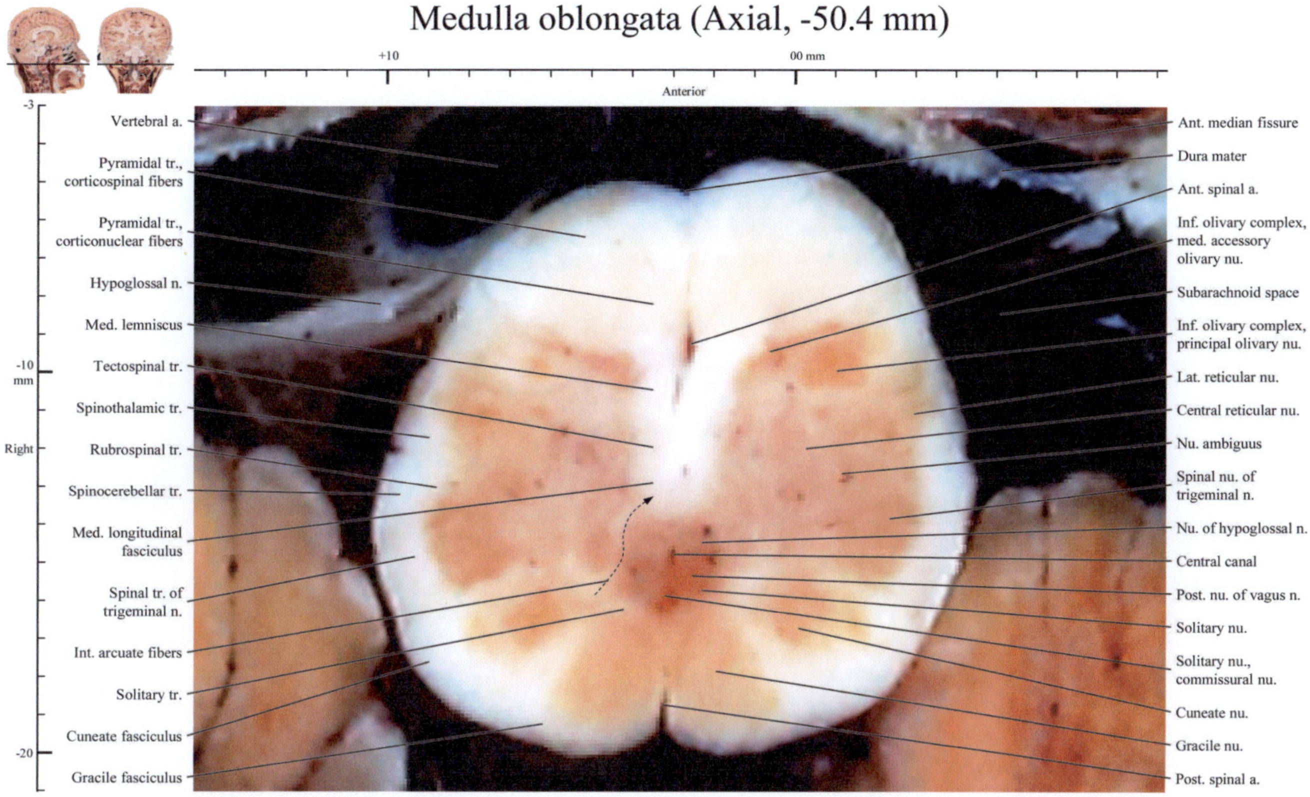

Medulla oblongata (Axial, -51.0 mm)

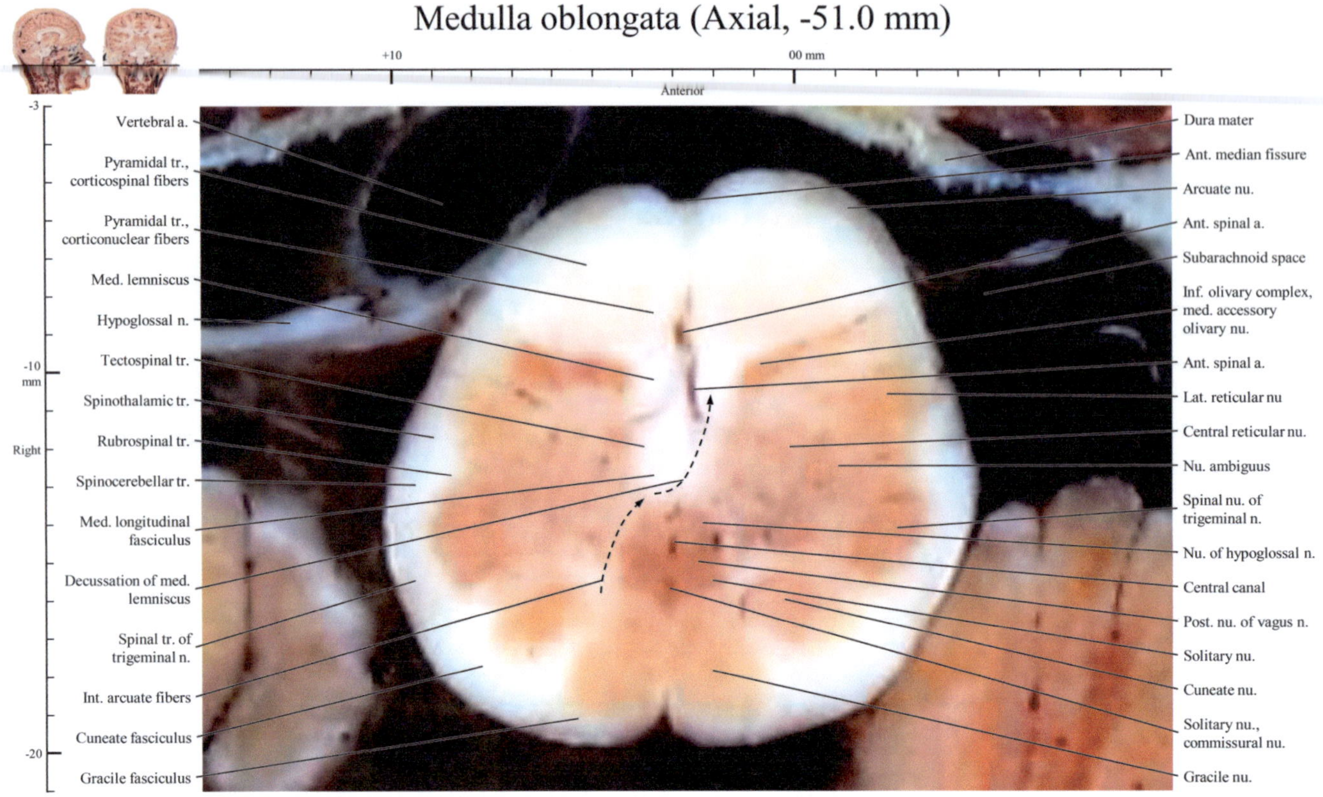

Medulla oblongata (Axial, -51.4 mm)

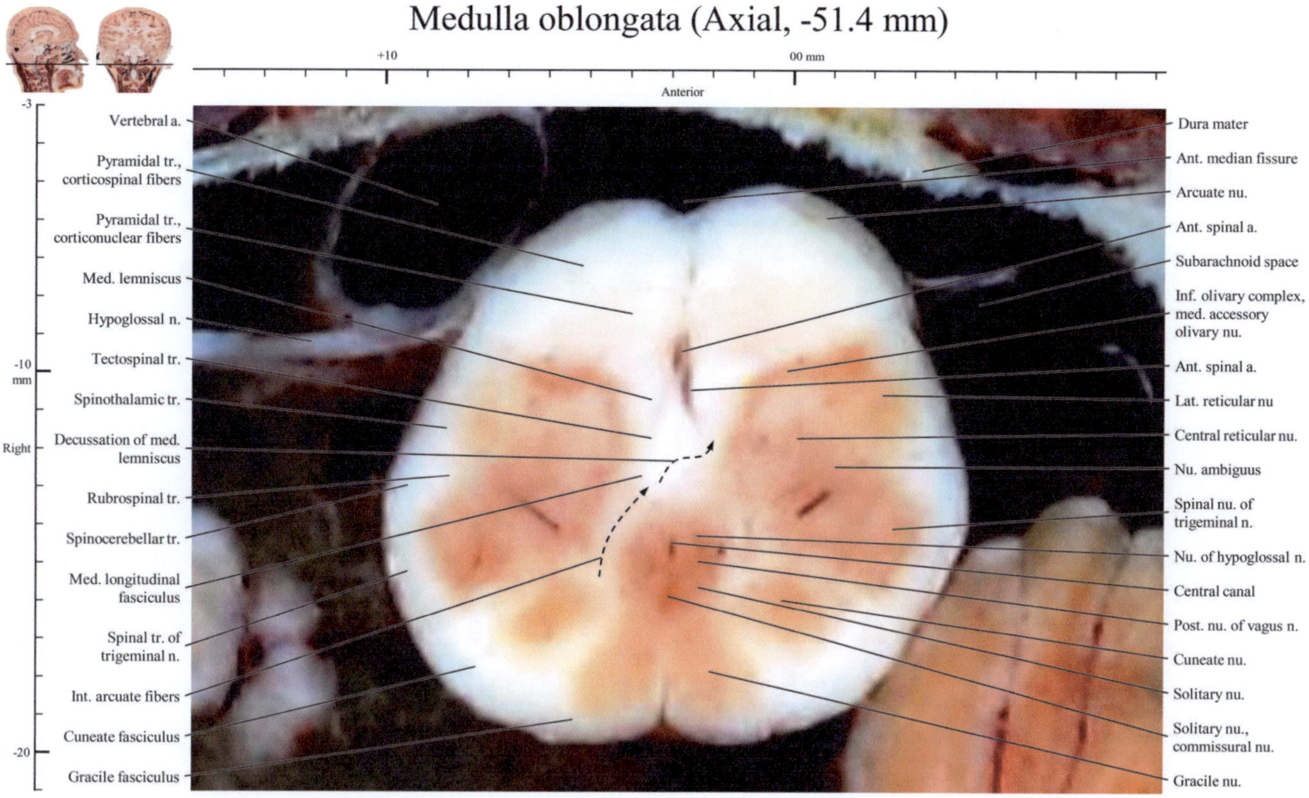

Medulla oblongata (Axial, -52.0 mm)

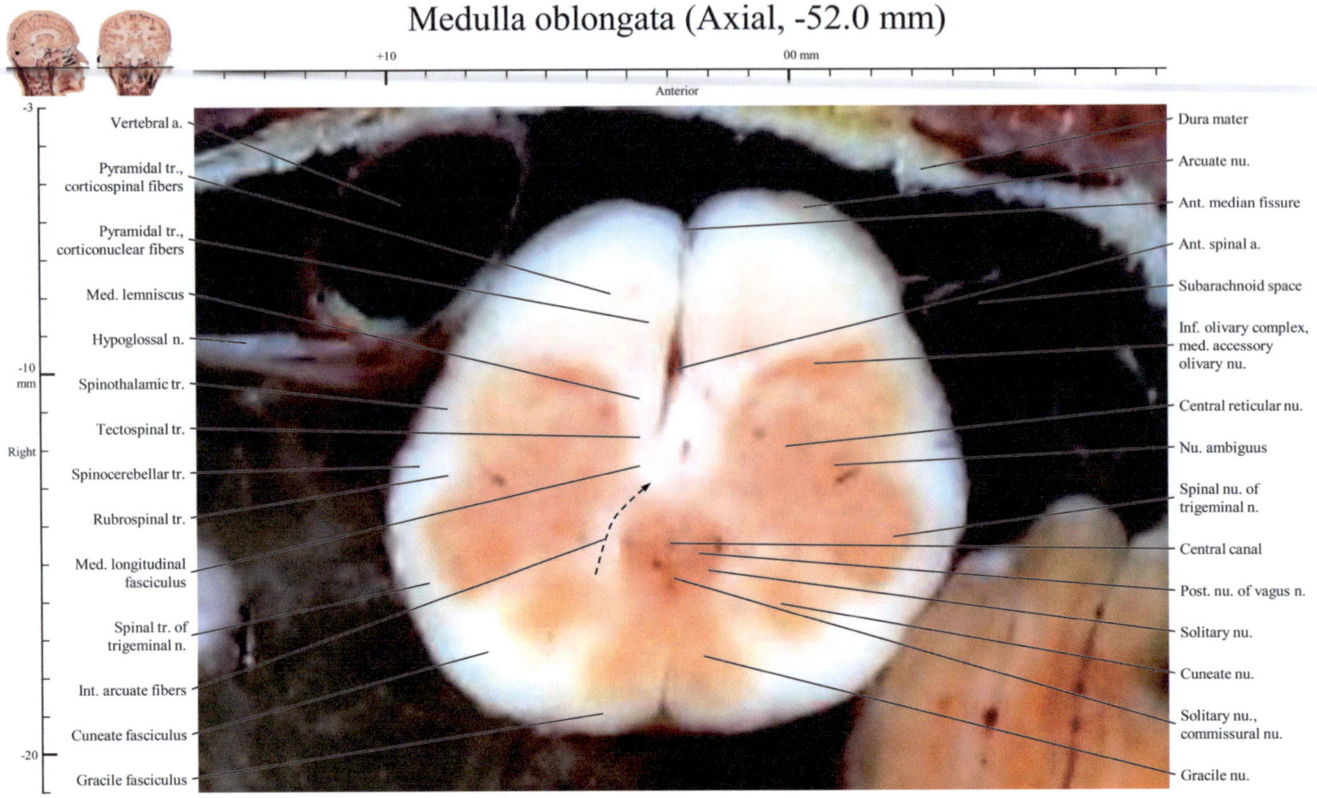

Medulla oblongata (Axial, -52.4 mm)

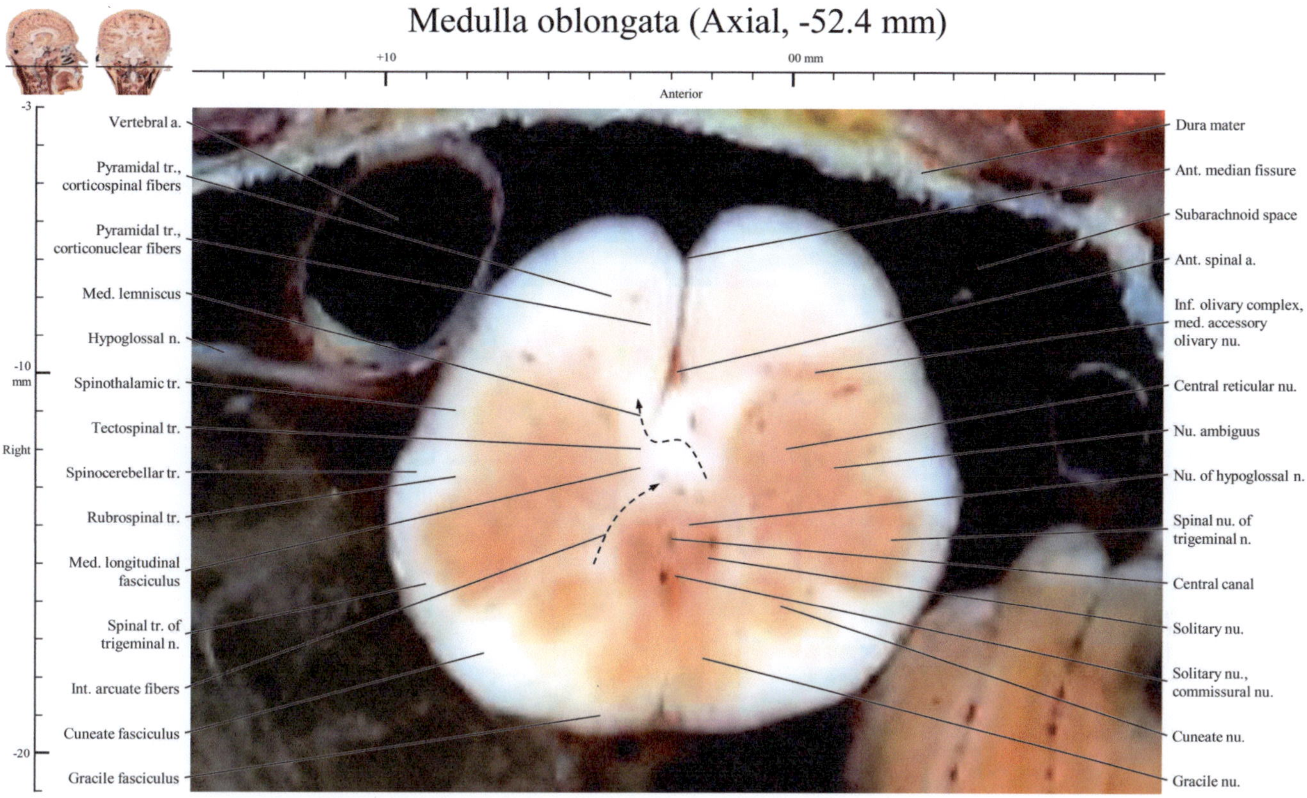

Medulla oblongata (Axial, -53.0 mm)

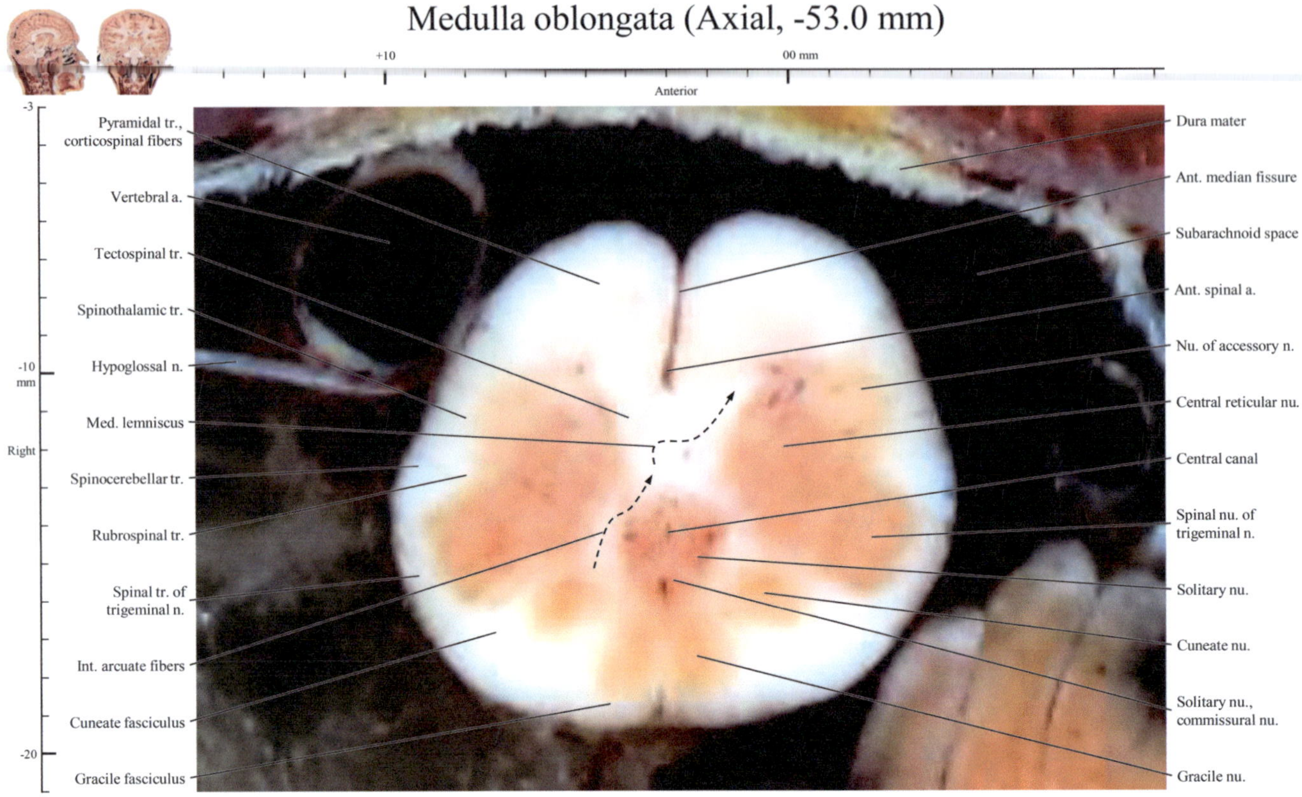

Medulla oblongata (Axial, -53.4 mm)

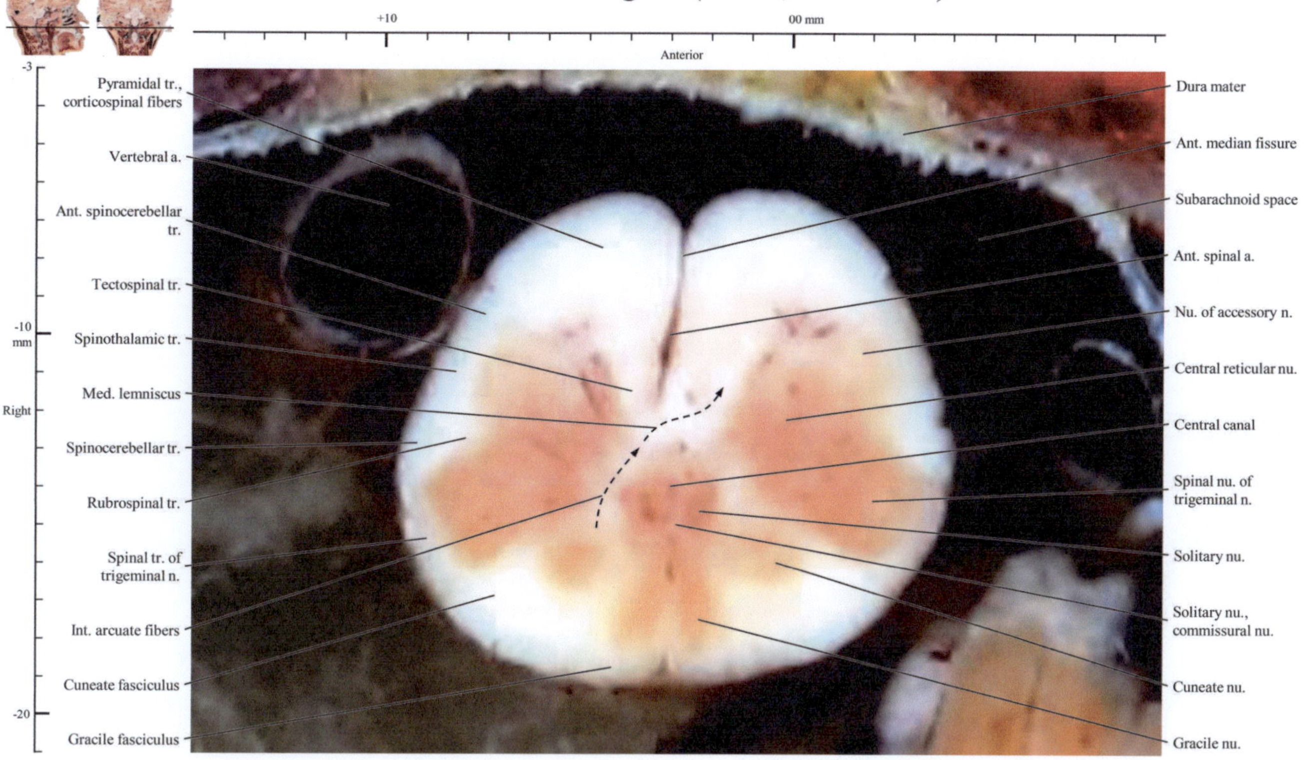

Medulla oblongata (Axial, -54.0 mm)

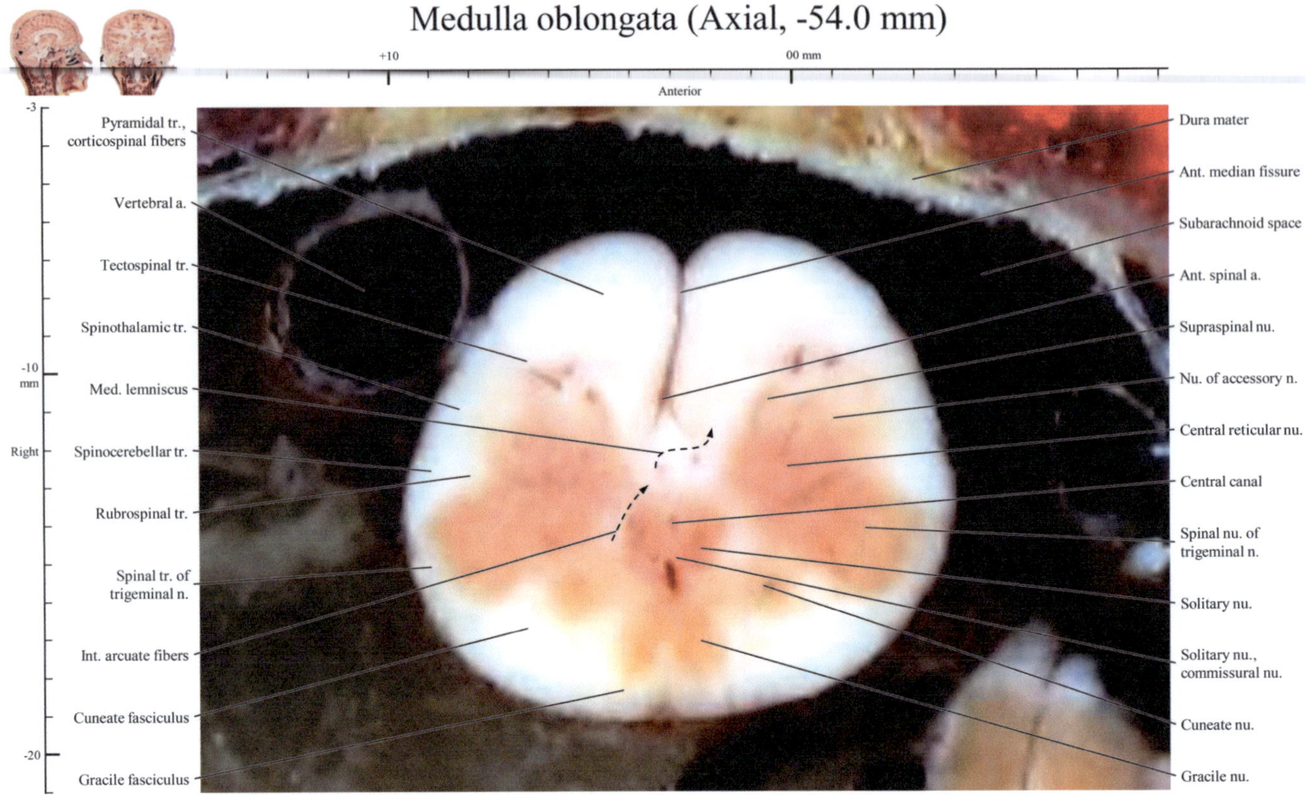

Medulla oblongata (Axial, -54.4 mm)

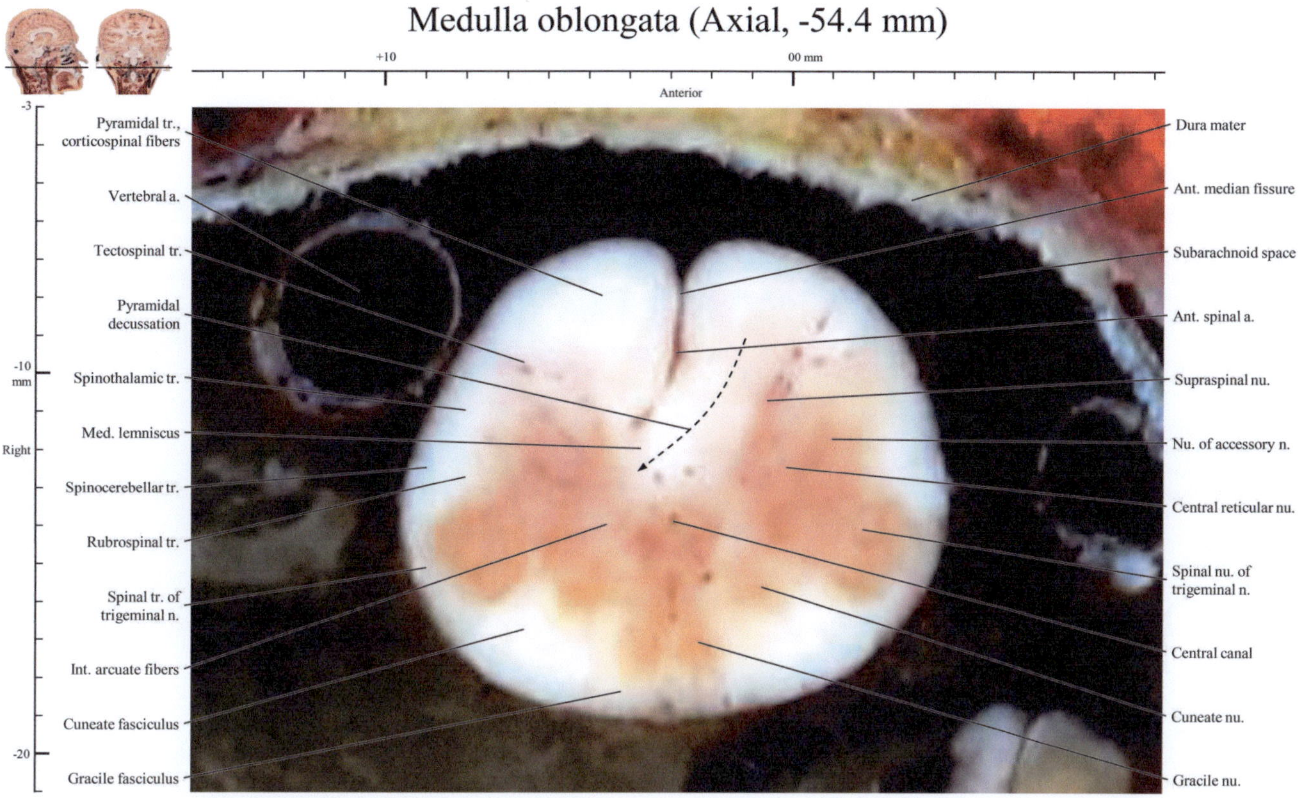

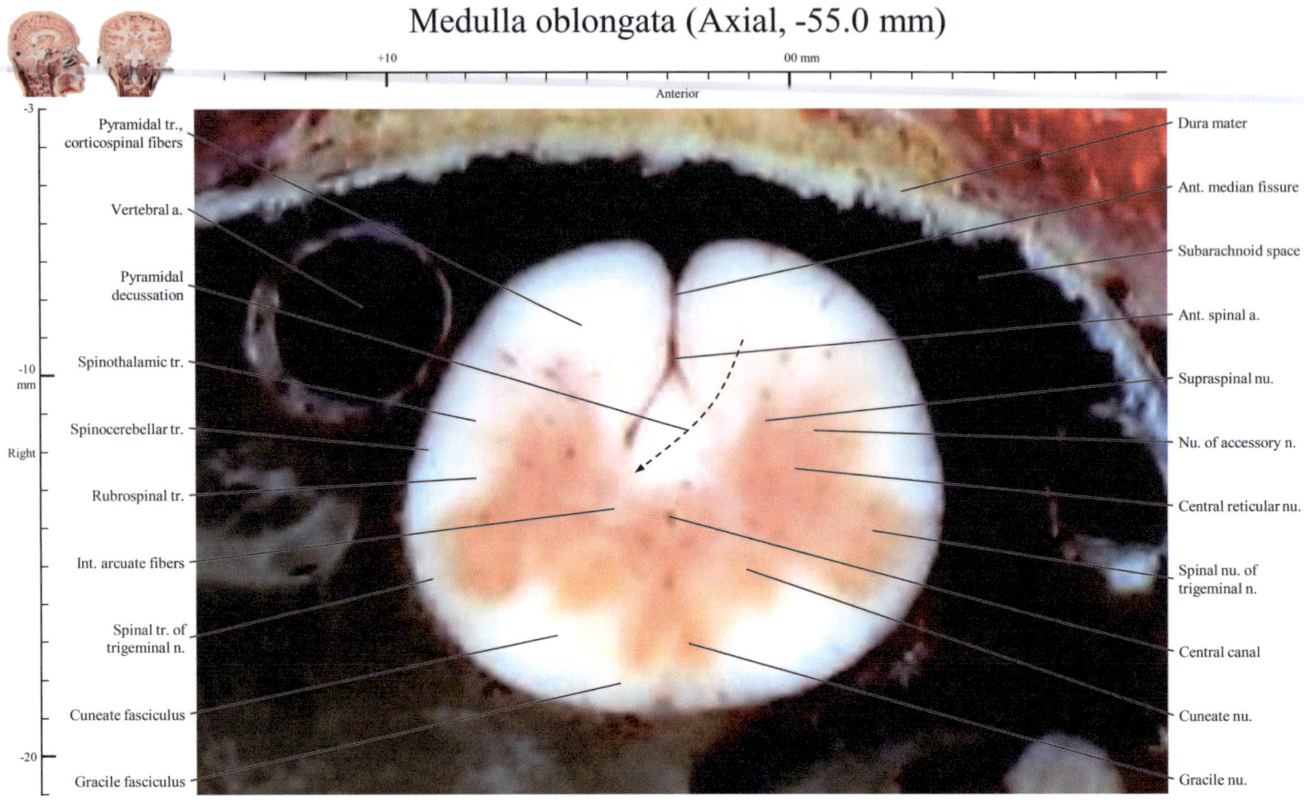
Medulla oblongata (Axial, -55.0 mm)
+10
00 mm
Anterior
-3
-10 mm
Right
-20
Pyramidal tr., corticospinal fibers
Vertebral a.
Pyramidal decussation
Spinothalamic tr.
Spinocerebellar tr.
Rubrospinal tr.
Int. arcuate fibers
Spinal tr. of trigeminal n.
Cuneate fasciculus
Gracile fasciculus
Dura mater
Ant. median fissure
Subarachnoid space
Ant. spinal a.
Supraspinal nu.
Nu. of accessory n.
Central reticular nu.
Spinal nu. of trigeminal n.
Central canal
Cuneate nu.
Gracile nu.

Medulla oblongata (Axial, -55.4 mm)

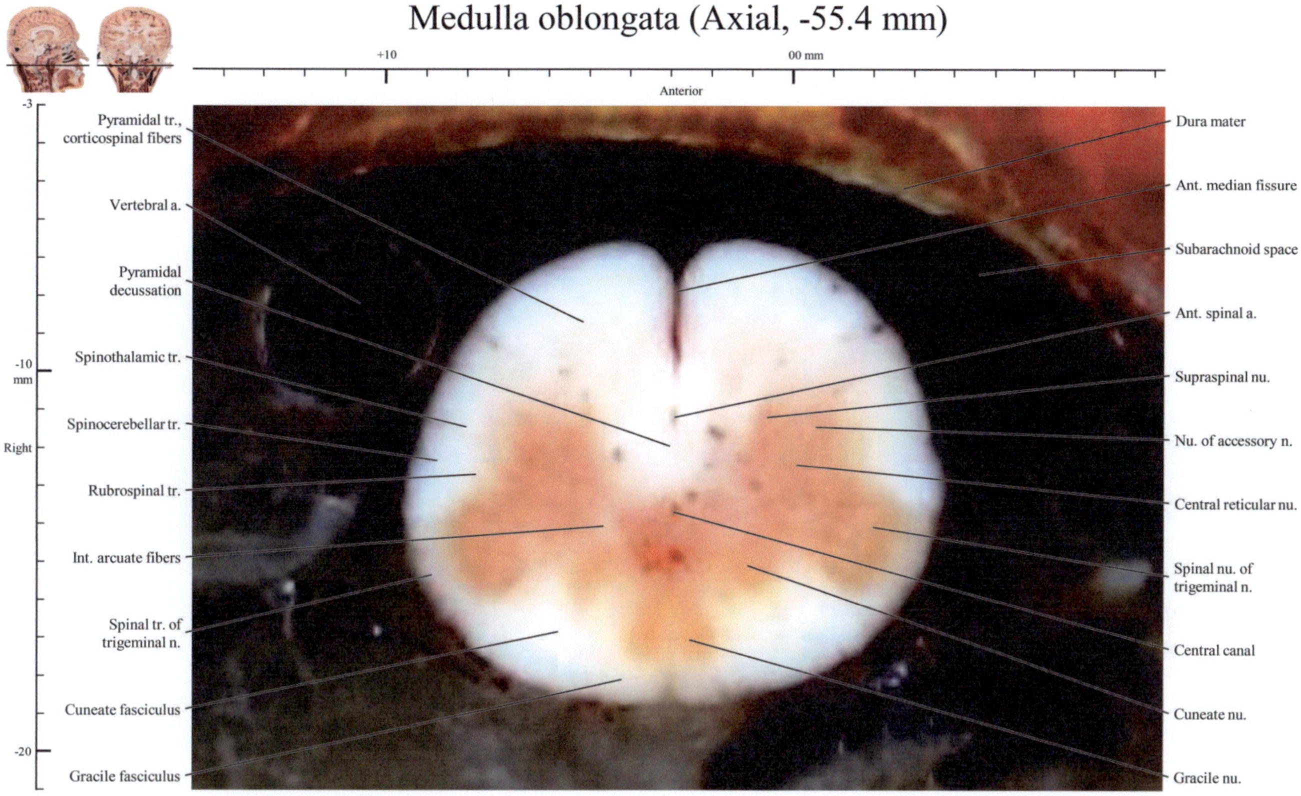

Medulla oblongata (Axial, -56.0 mm)

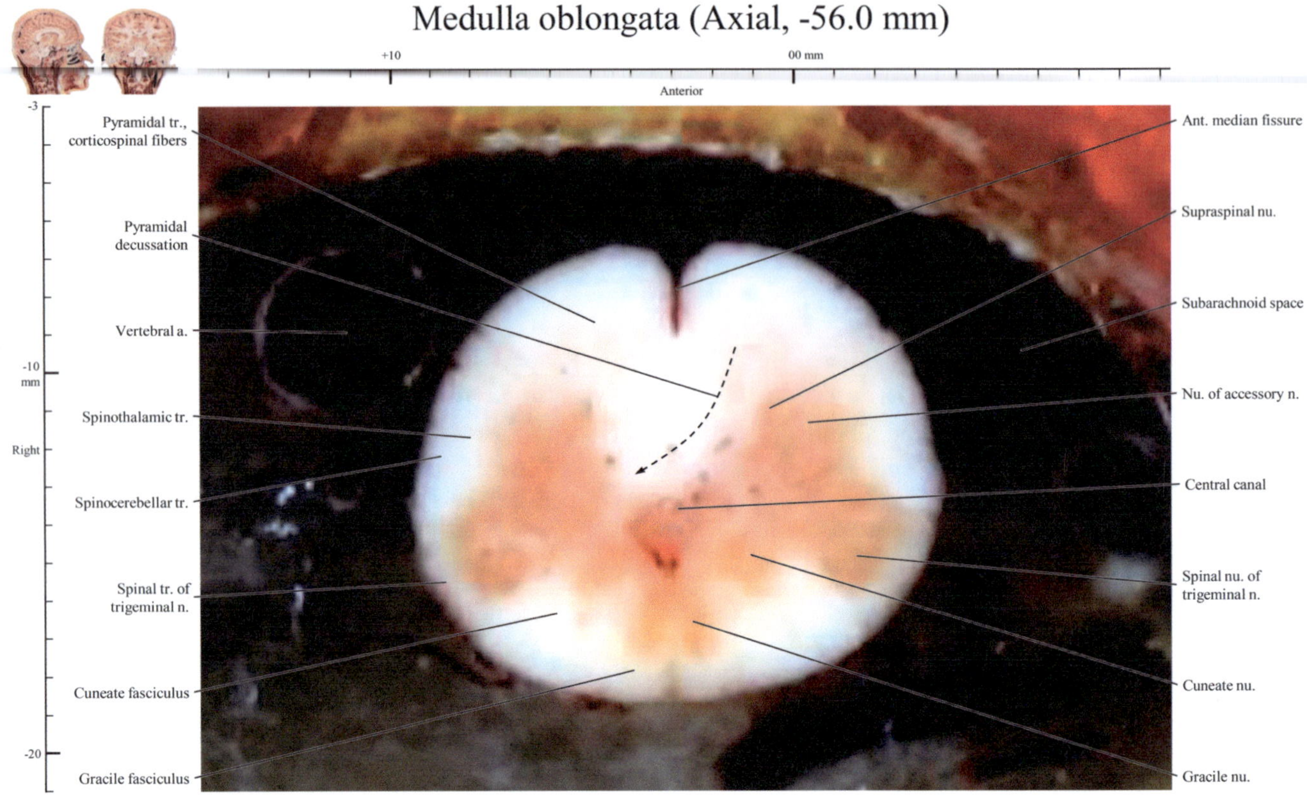

Medulla oblongata (Axial, -56.4 mm)

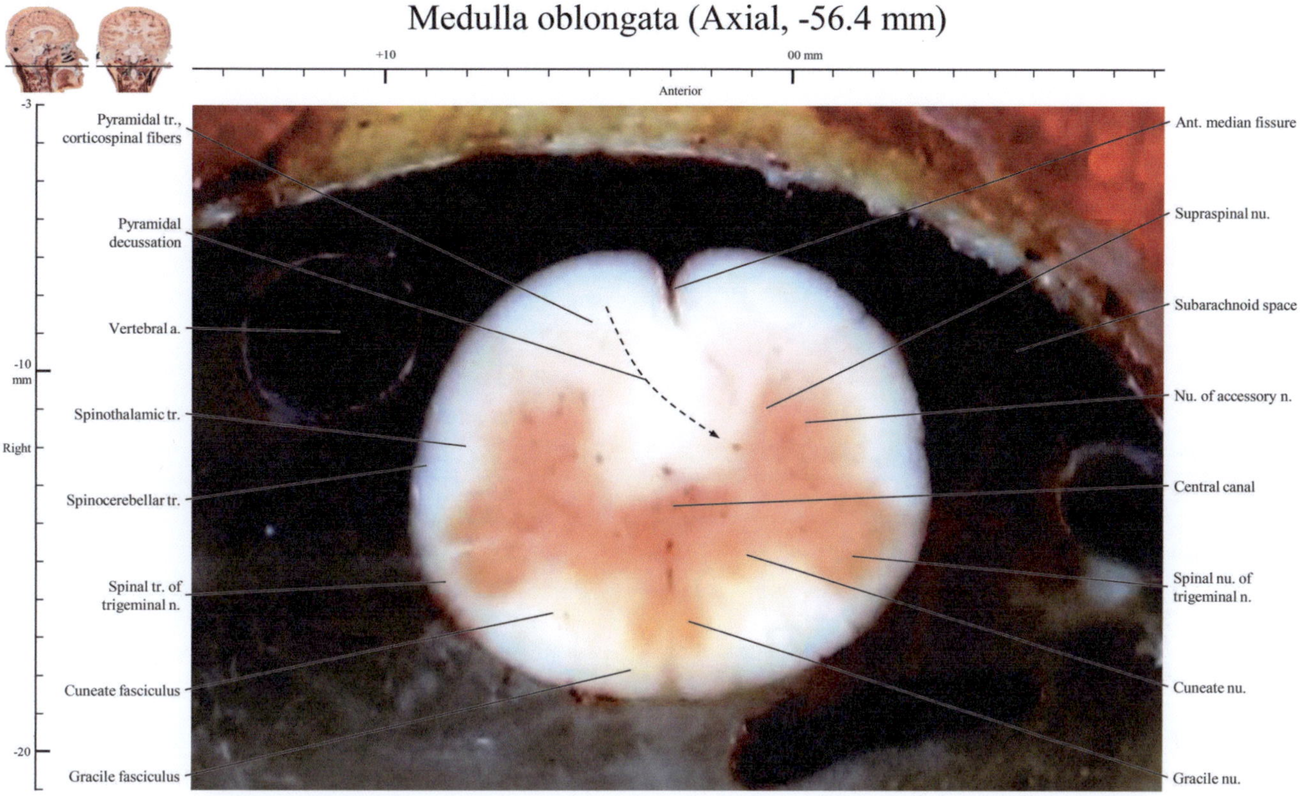

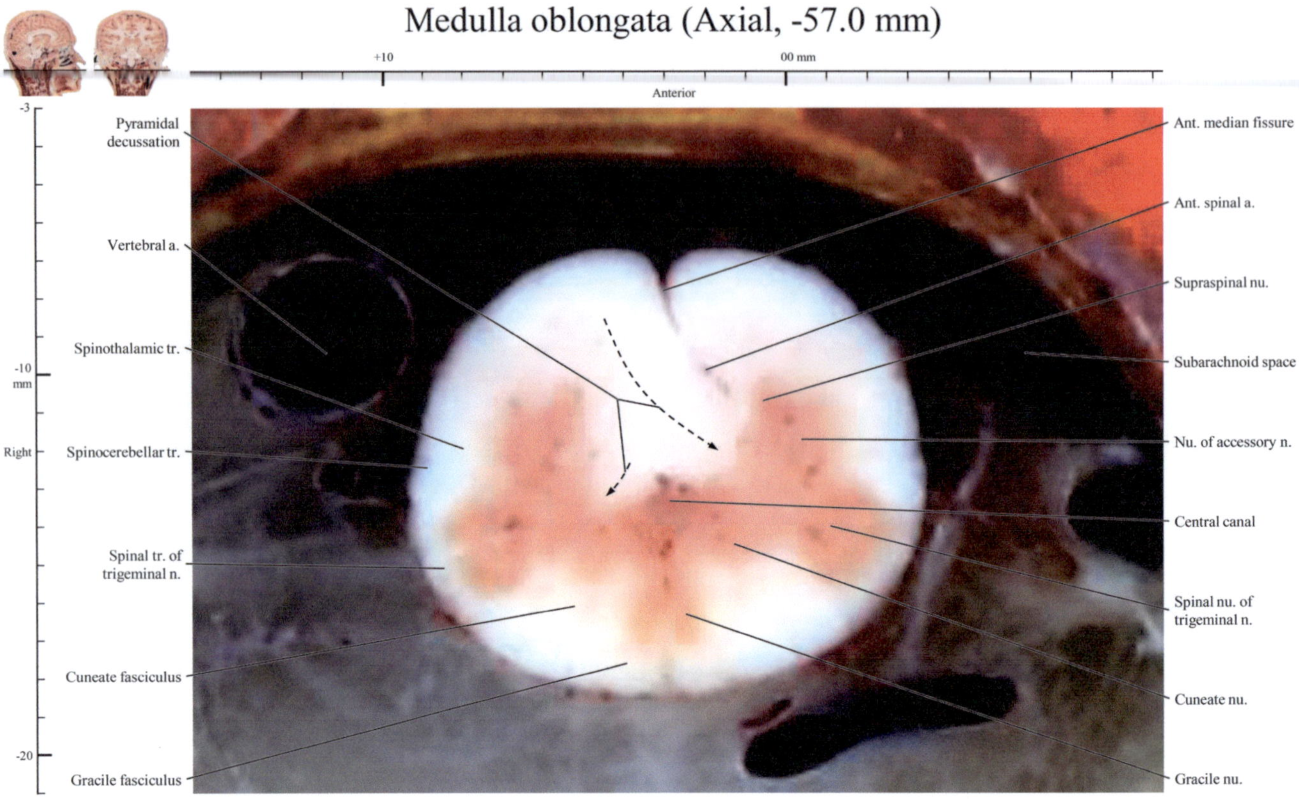
Medulla oblongata (Axial, -57.0 mm)
+10
00 mm
Anterior
-3
-10 mm
Right
-20
Pyramidal decussation
Vertebral a.
Spinothalamic tr.
Spinocerebellar tr.
Spinal tr. of trigeminal n.
Cuneate fasciculus
Gracile fasciculus
Ant. median fissure
Ant. spinal a.
Supraspinal nu.
Subarachnoid space
Nu. of accessory n.
Central canal
Spinal nu. of trigeminal n.
Cuneate nu.
Gracile nu.

Medulla oblongata (Axial, -57.4 mm)

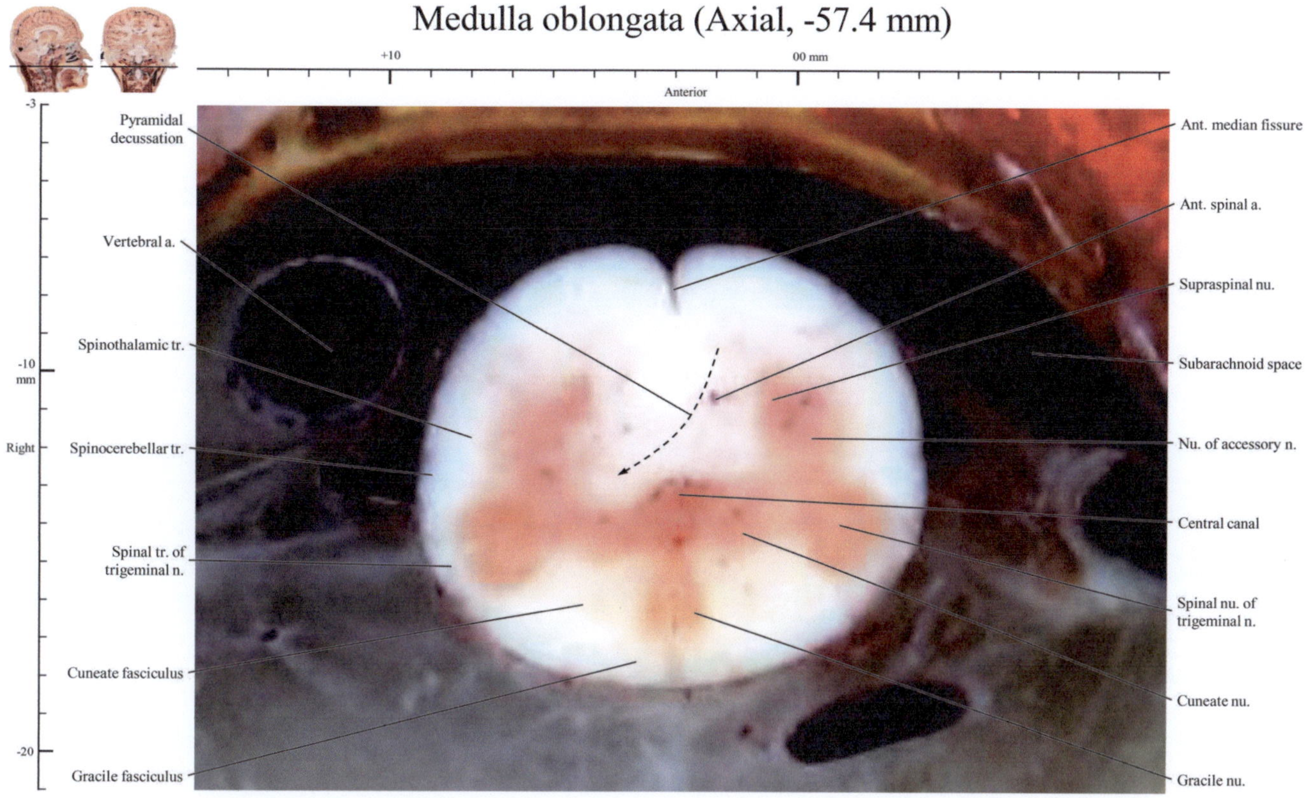

Spinal cord (Axial, -58.0 mm)

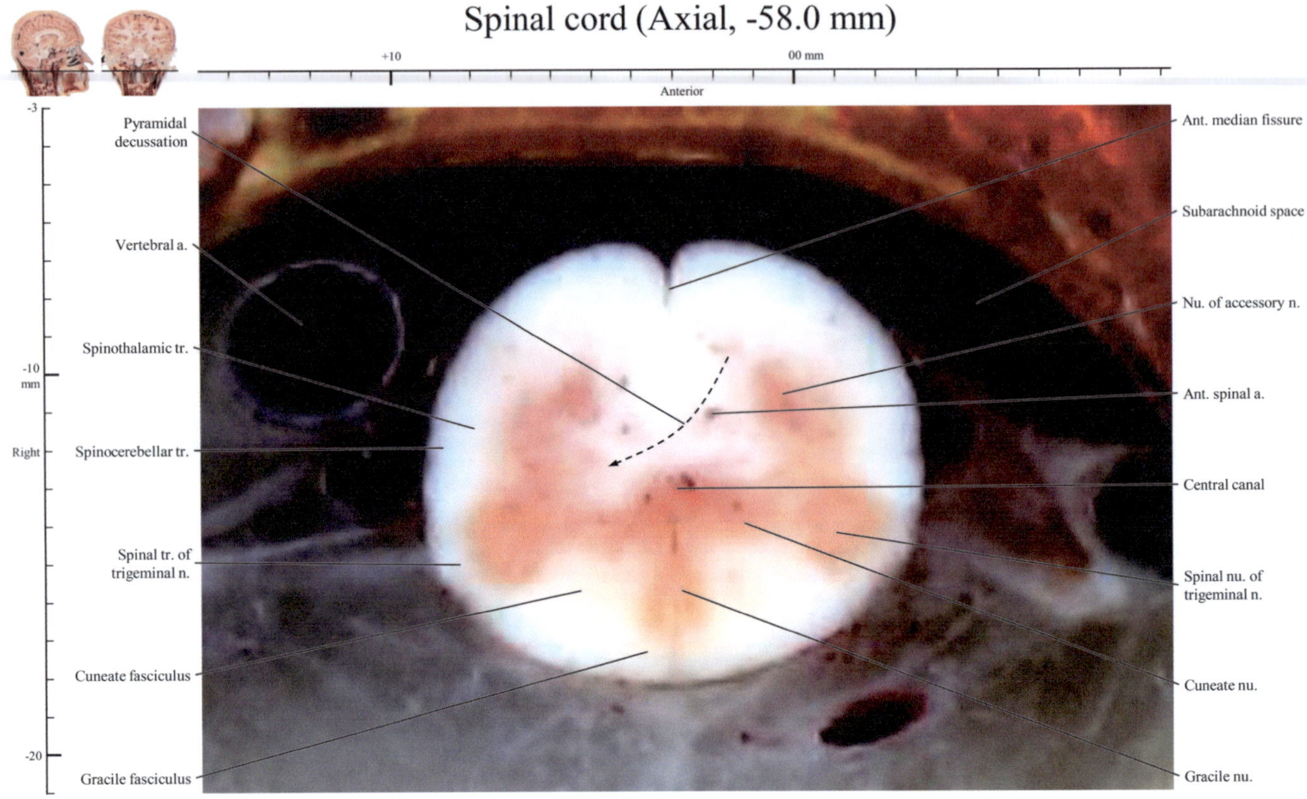

3 Coronal Brainstem

Images of +9.0 mm ~ +6.3 mm, Pons

Images of +5.4 mm ~ −3.6 mm, Midbrain, pons

Images of −4.5 mm ~ −18.0 mm, Midbrain, pons, medulla oblongata

Images of −18.9 mm ~ 21.6 mm, Midbrain, medulla oblongata

Images of −22.5 mm, Midbrain

All images, 0.06 mm × 0.06 mm sized-pixel and 0.9 mm intervals

Abbreviation: a., artery; ant., anterior; inf., inferior; int., internal; lat., lateral; m., muscle; med., medial; n., nerve; nu., nucleus; post., posterior; sup., superior; tr., tract

J. S. Park, Y. You, *Cross-Sectional Atlas of Human Brainstem*, https://doi.org/10.1007/978-981-19-9519-4_3

Pons (Coronal, +9.0 mm)

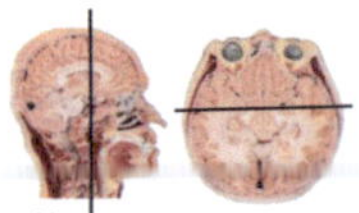

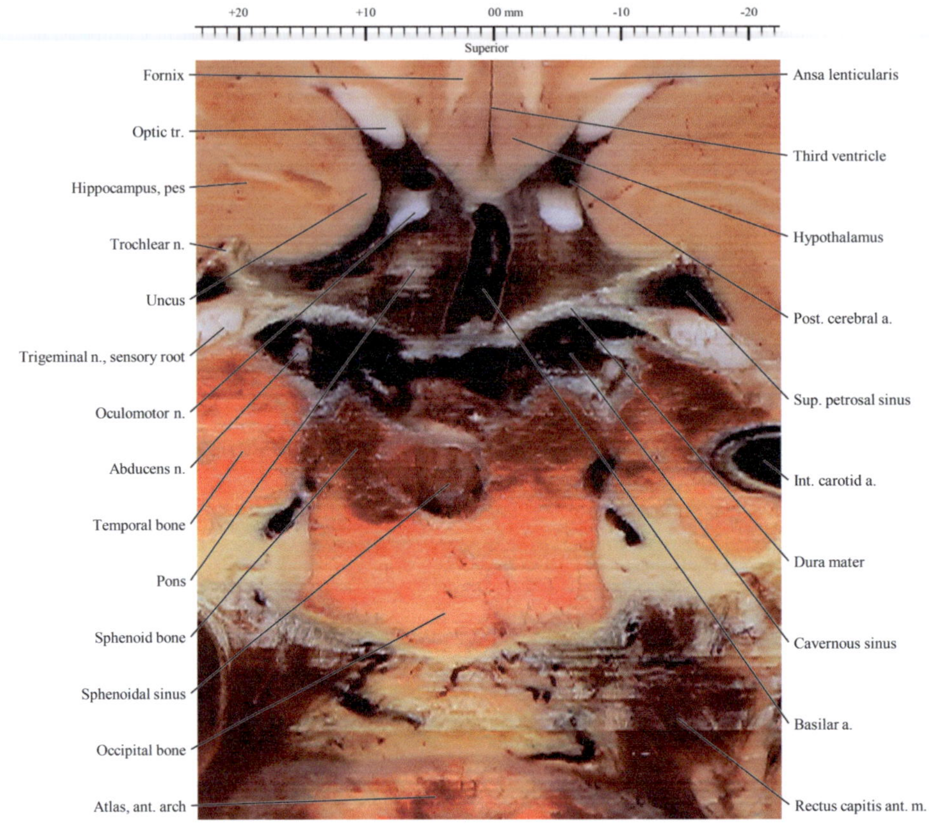

Pons (Coronal, +8.1 mm)

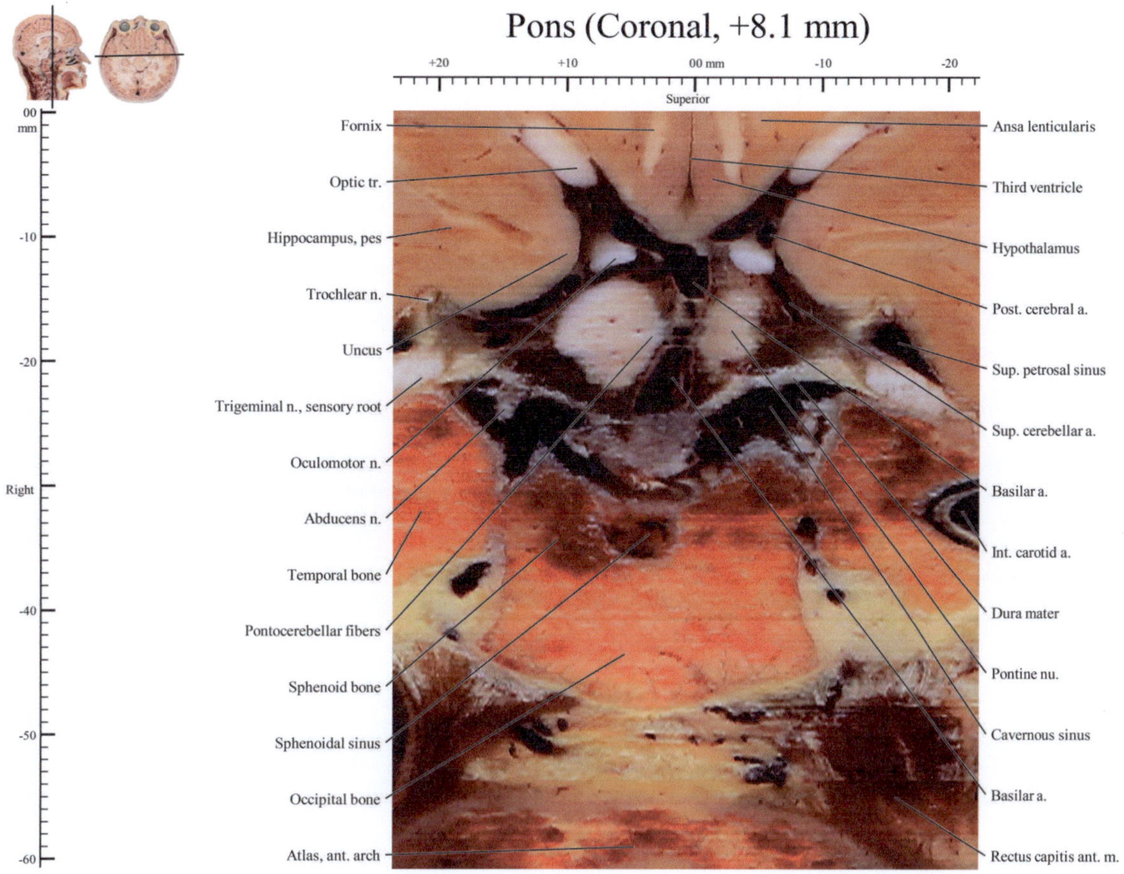

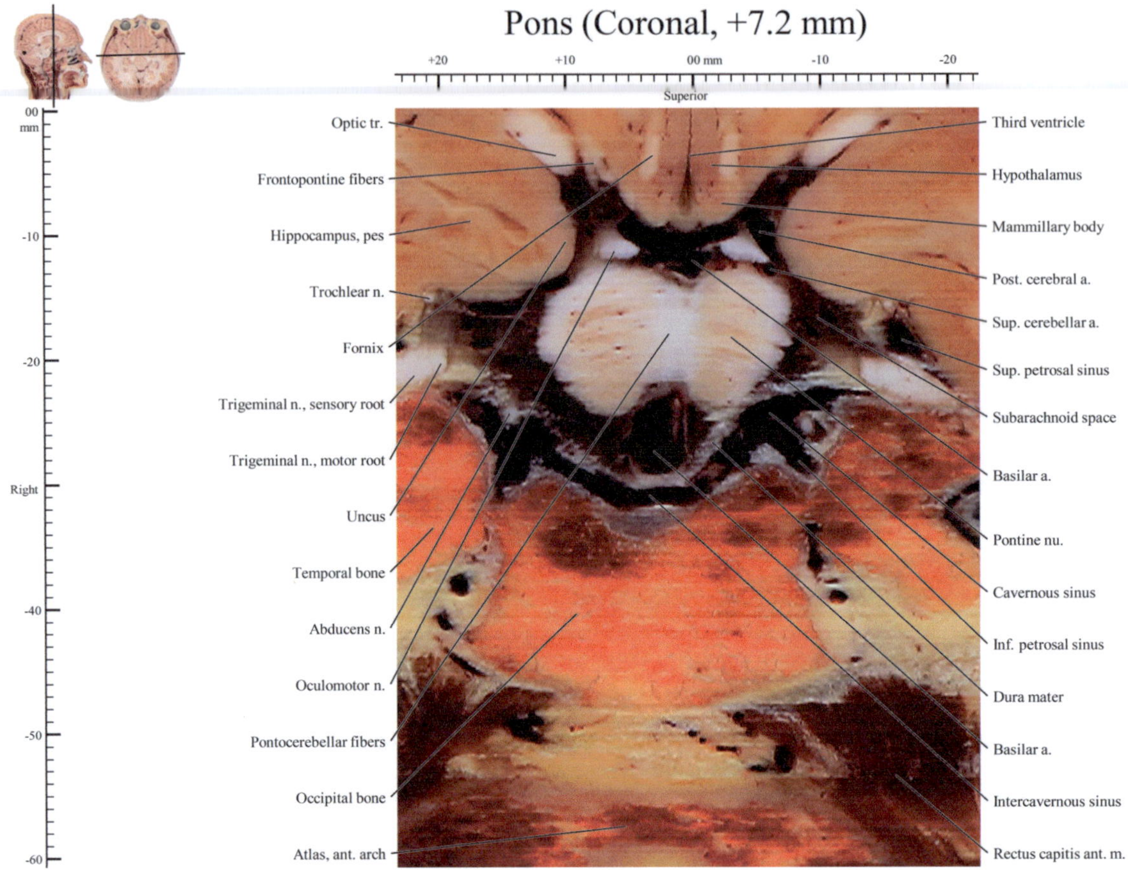
Pons (Coronal, +7.2 mm)
+20
+10
00 mm
-10
-20
Superior
00 mm
-10
-20
Right
-40
-50
-60
Optic tr.
Frontopontine fibers
Hippocampus, pes
Trochlear n.
Fornix
Trigeminal n., sensory root
Trigeminal n., motor root
Uncus
Temporal bone
Abducens n.
Oculomotor n.
Pontocerebellar fibers
Occipital bone
Atlas, ant. arch
Third ventricle
Hypothalamus
Mammillary body
Post. cerebral a.
Sup. cerebellar a.
Sup. petrosal sinus
Subarachnoid space
Basilar a.
Pontine nu.
Cavernous sinus
Inf. petrosal sinus
Dura mater
Basilar a.
Intercavernous sinus
Rectus capitis ant. m.

Pons (Coronal, +6.3 mm)

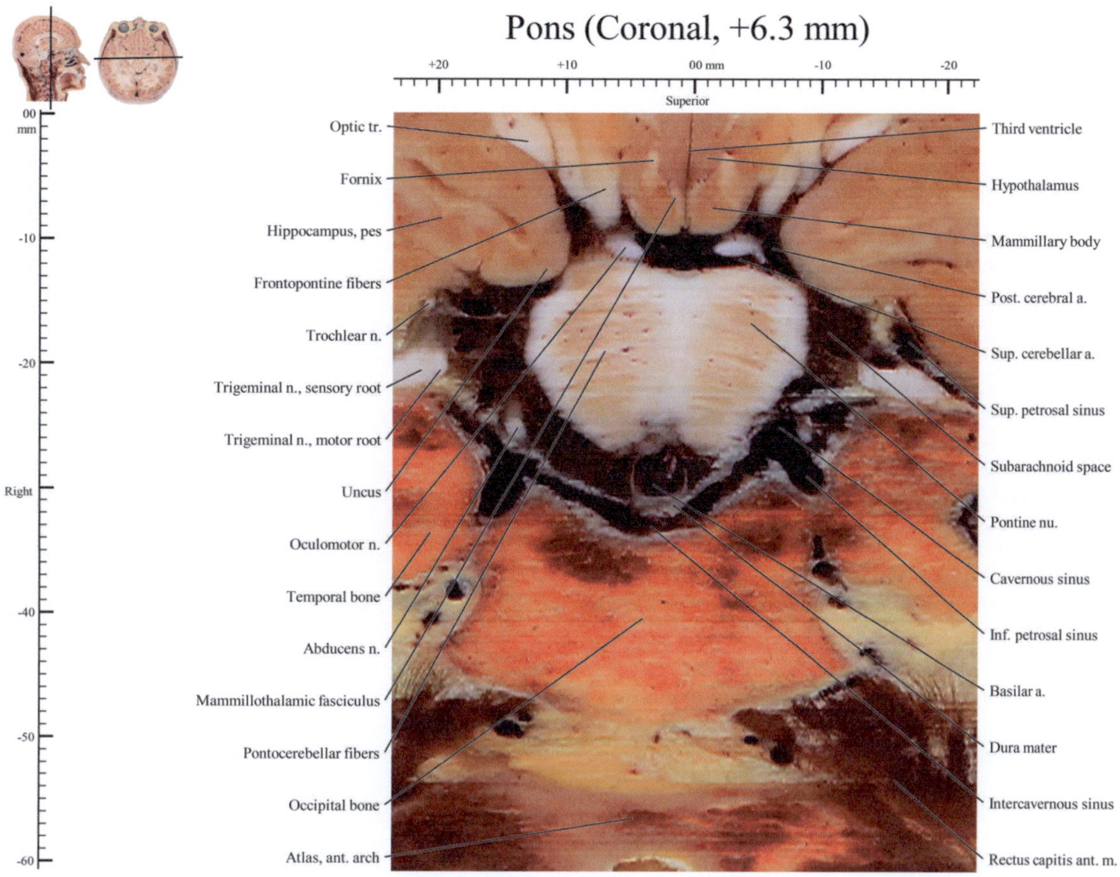

Midbrain, pons (Coronal, +5.4 mm)

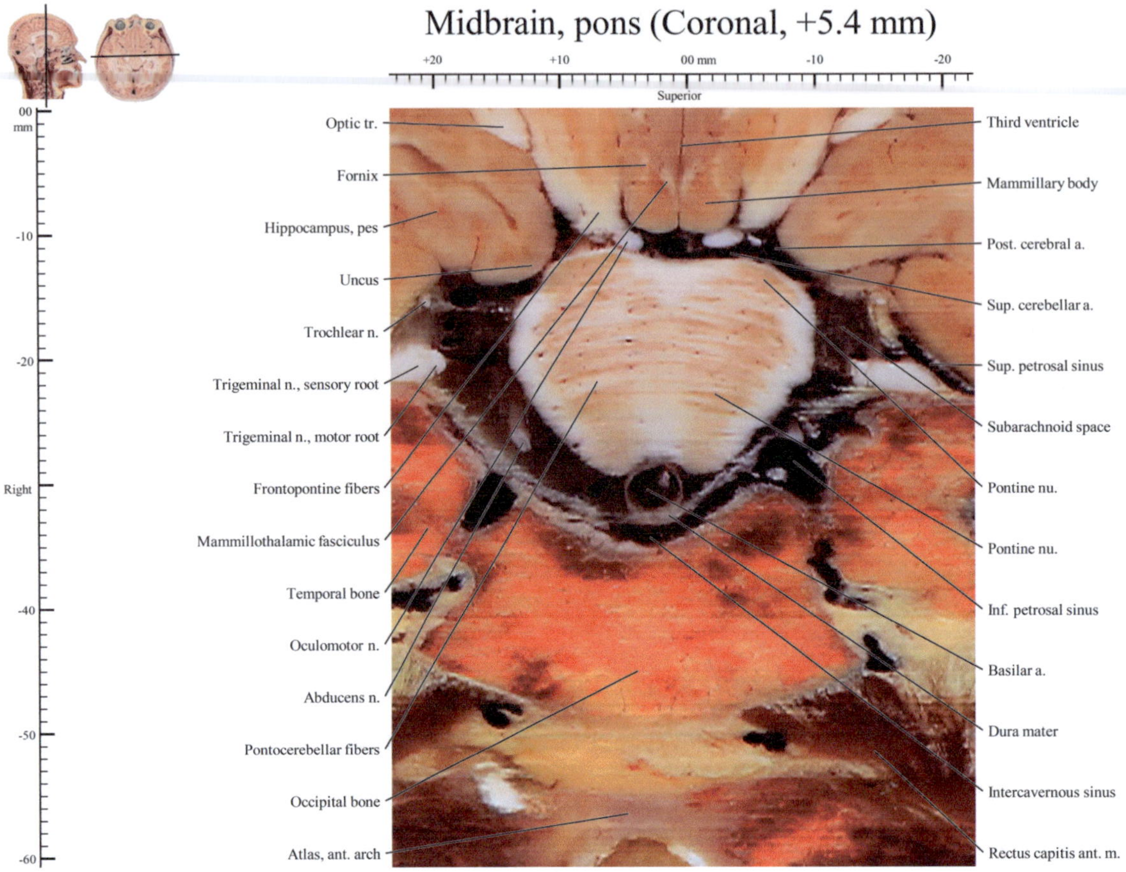

Midbrain, pons (Coronal, +4.5 mm)

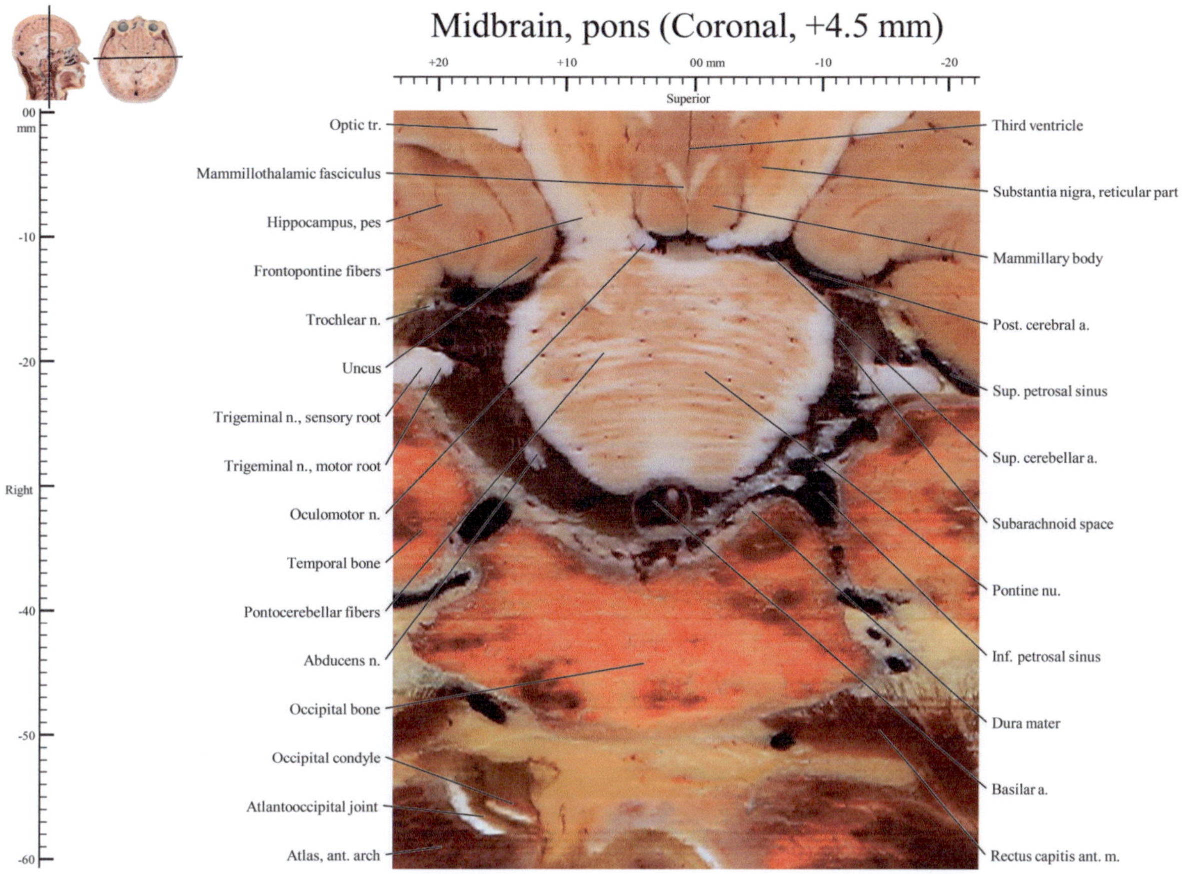

Midbrain, pons (Coronal, +3.6 mm)

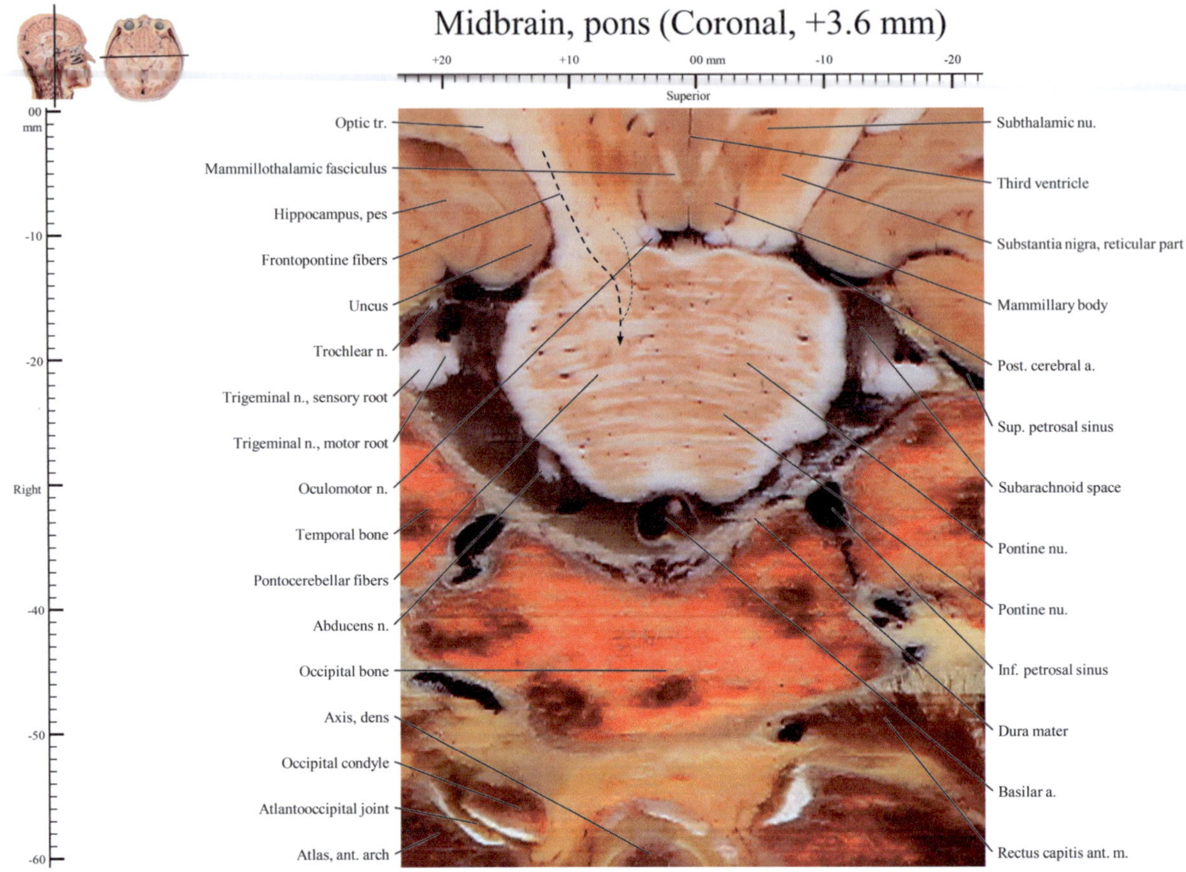

Midbrain, pons (Coronal, +2.7 mm)

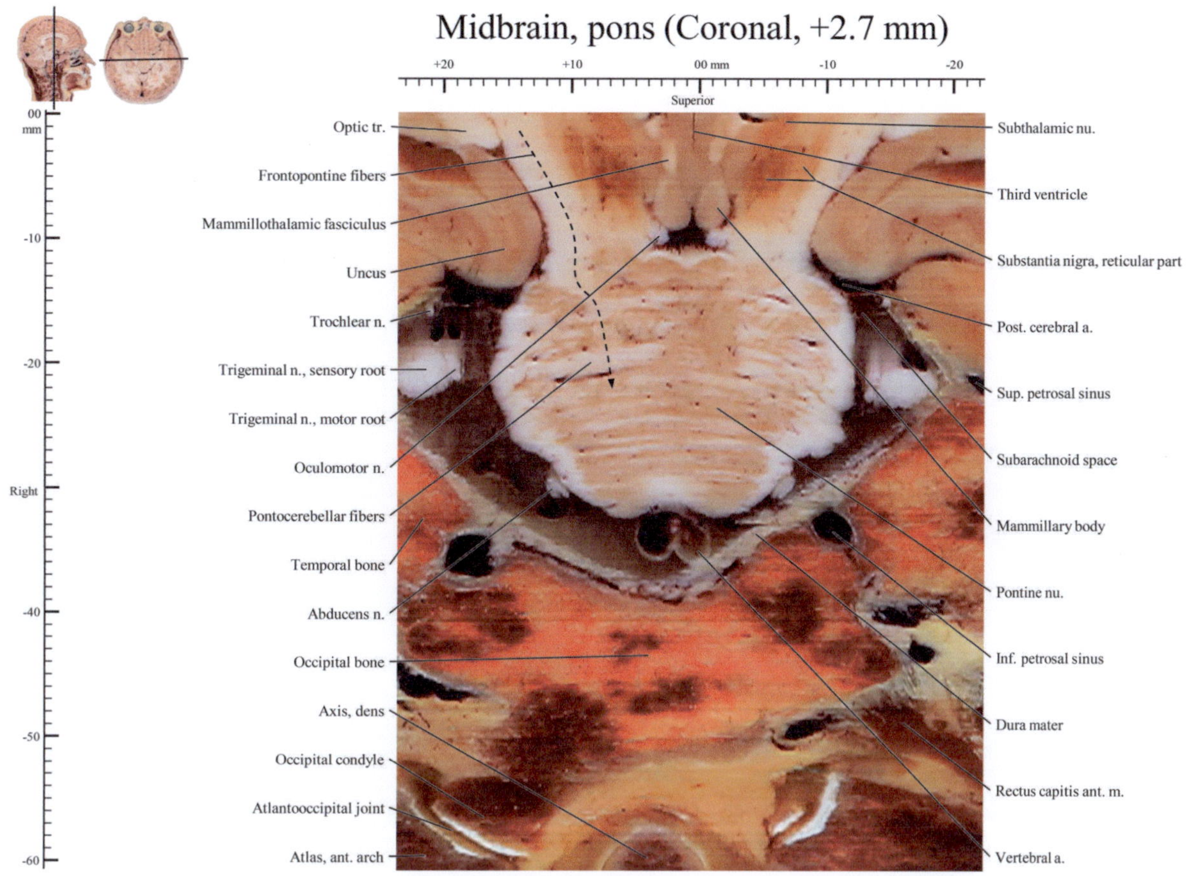

Midbrain, pons (Coronal, +1.8 mm)

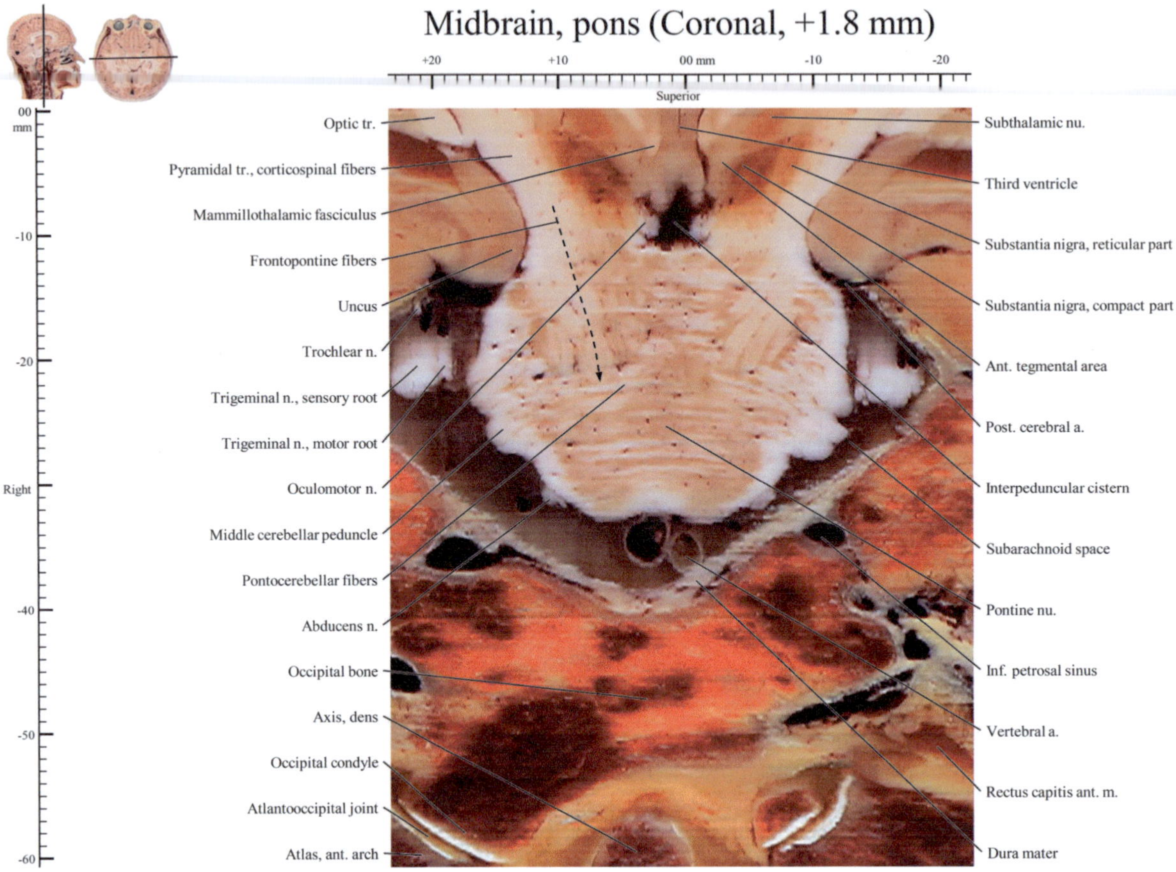

Midbrain, pons (Coronal, +0.9 mm)

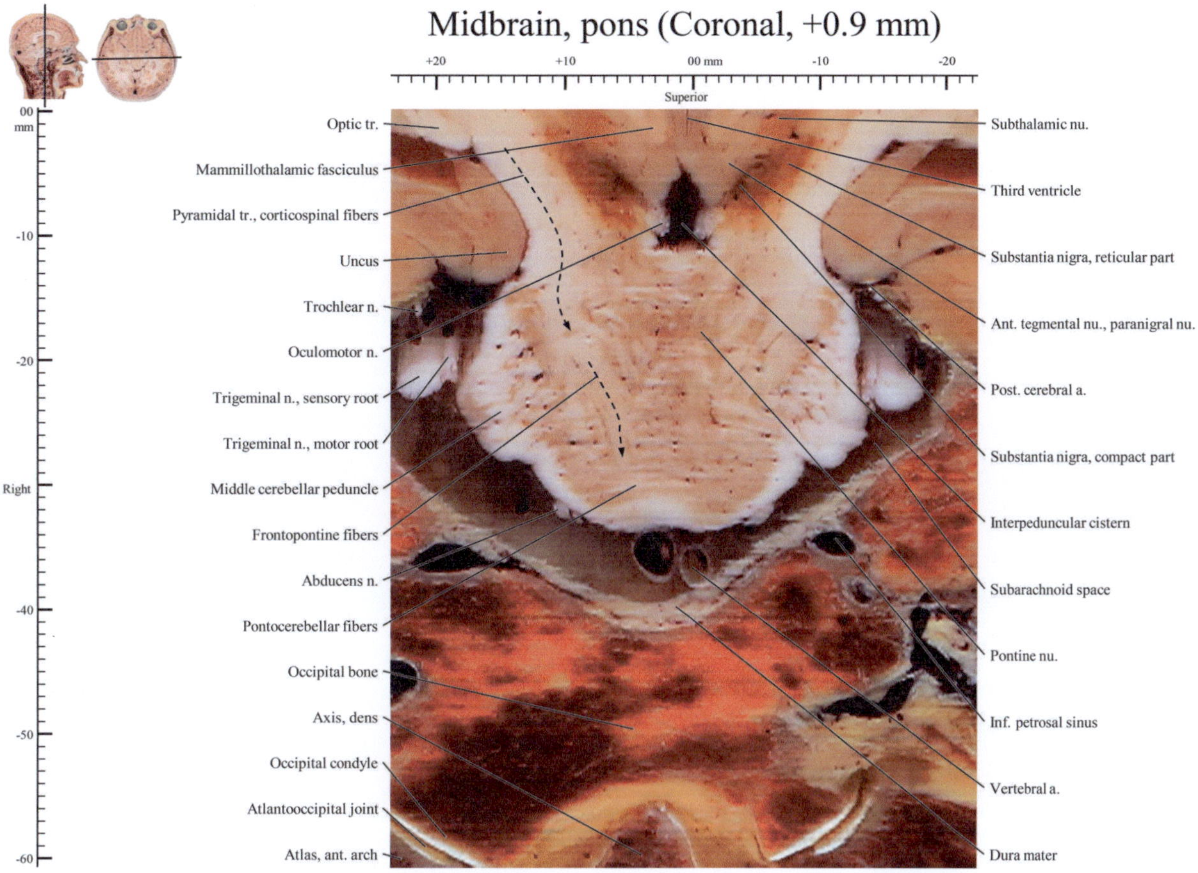

Midbrain, pons (Coronal, 0 mm)

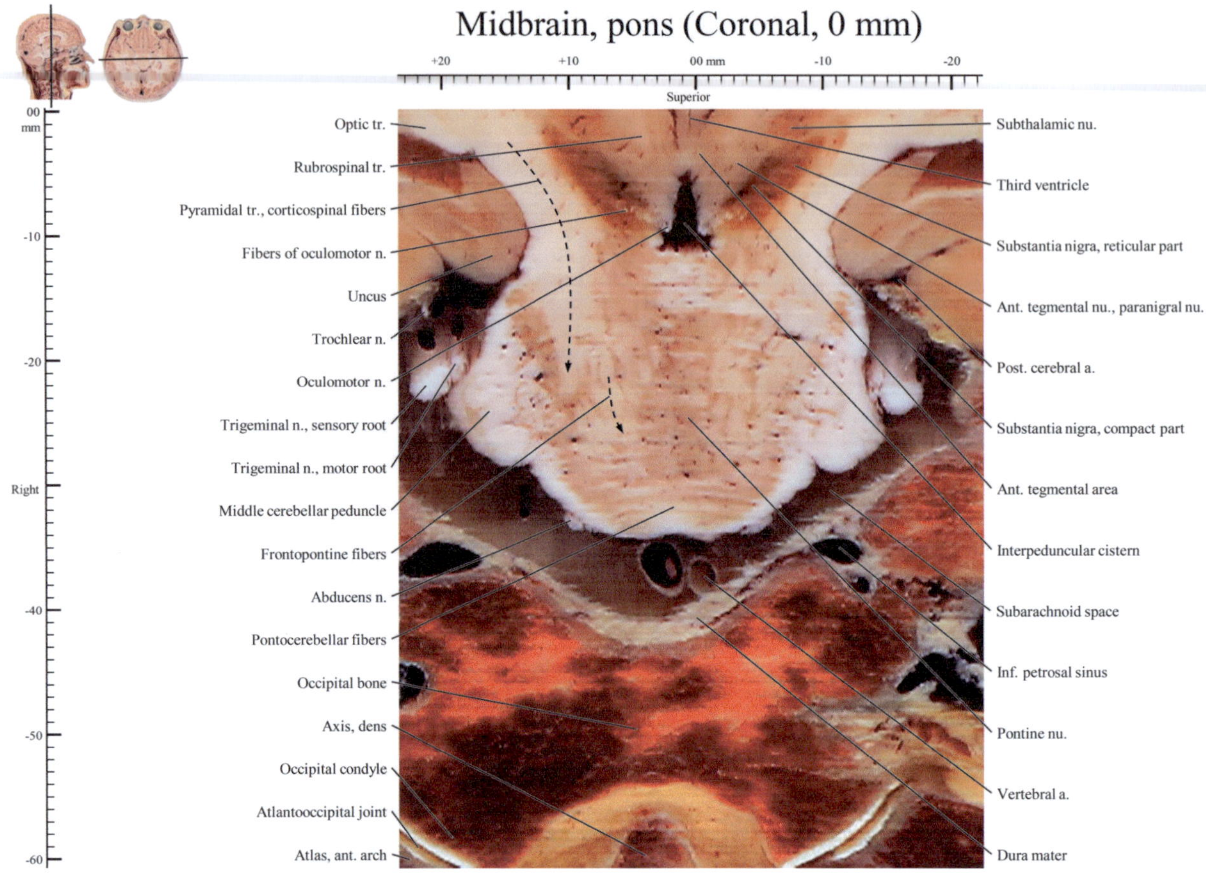

Midbrain, pons (Coronal, -0.9 mm)

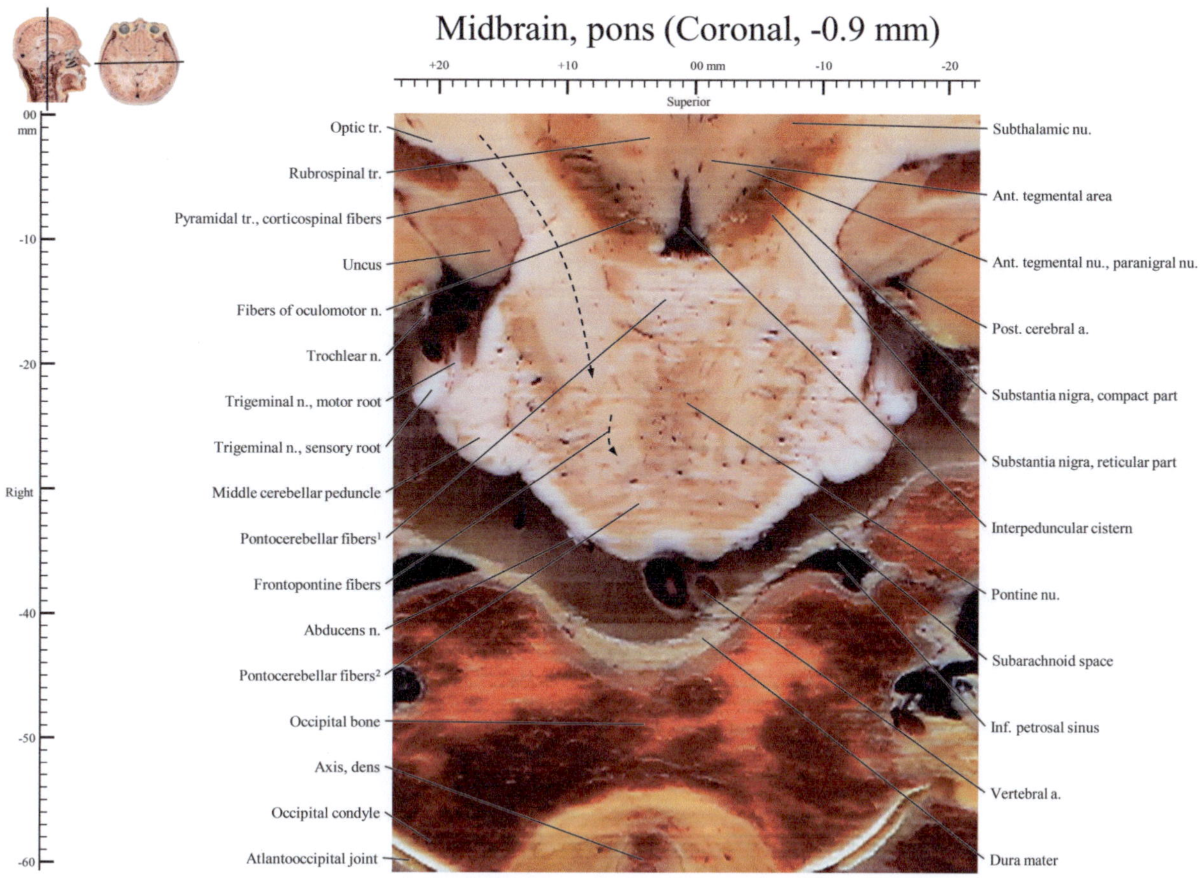

Midbrain, pons (Coronal, -1.8 mm)

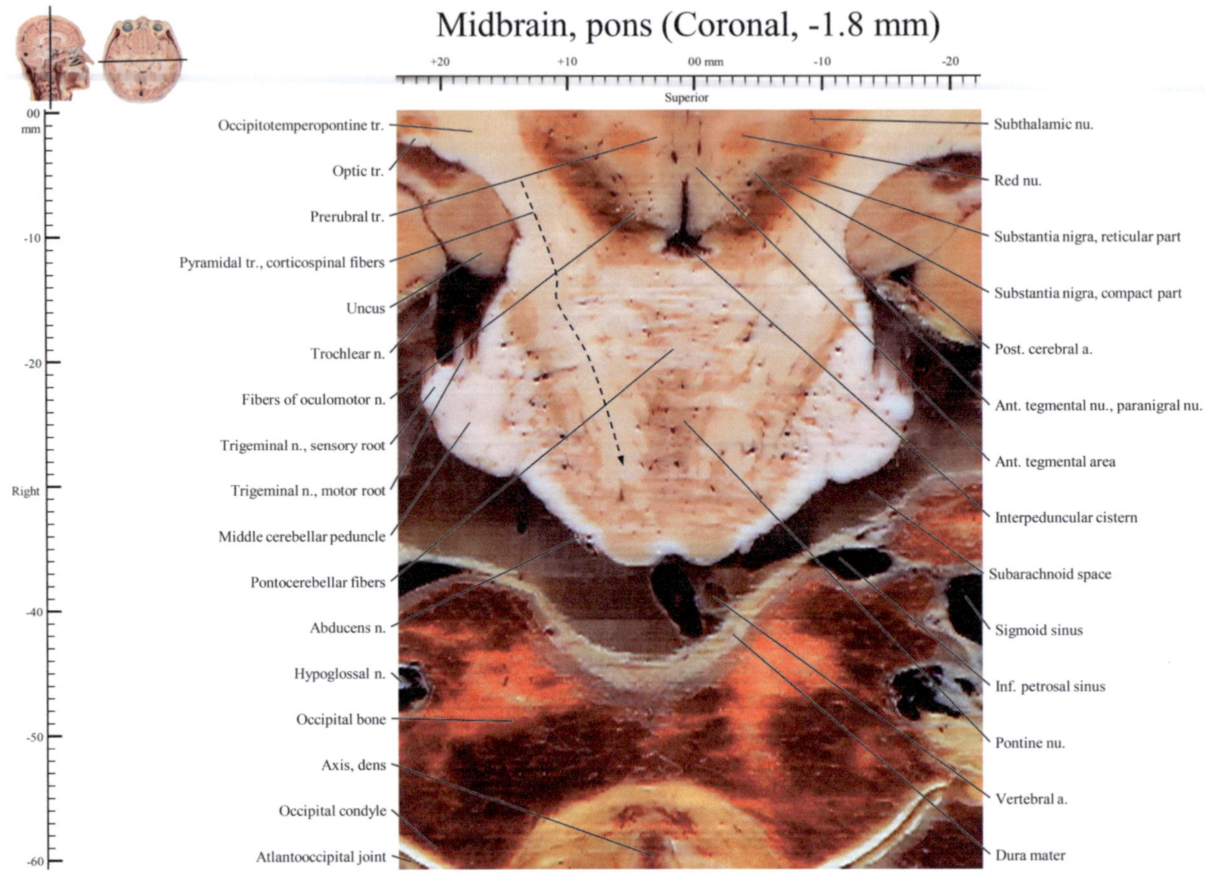

Midbrain, pons (Coronal, -2.7 mm)

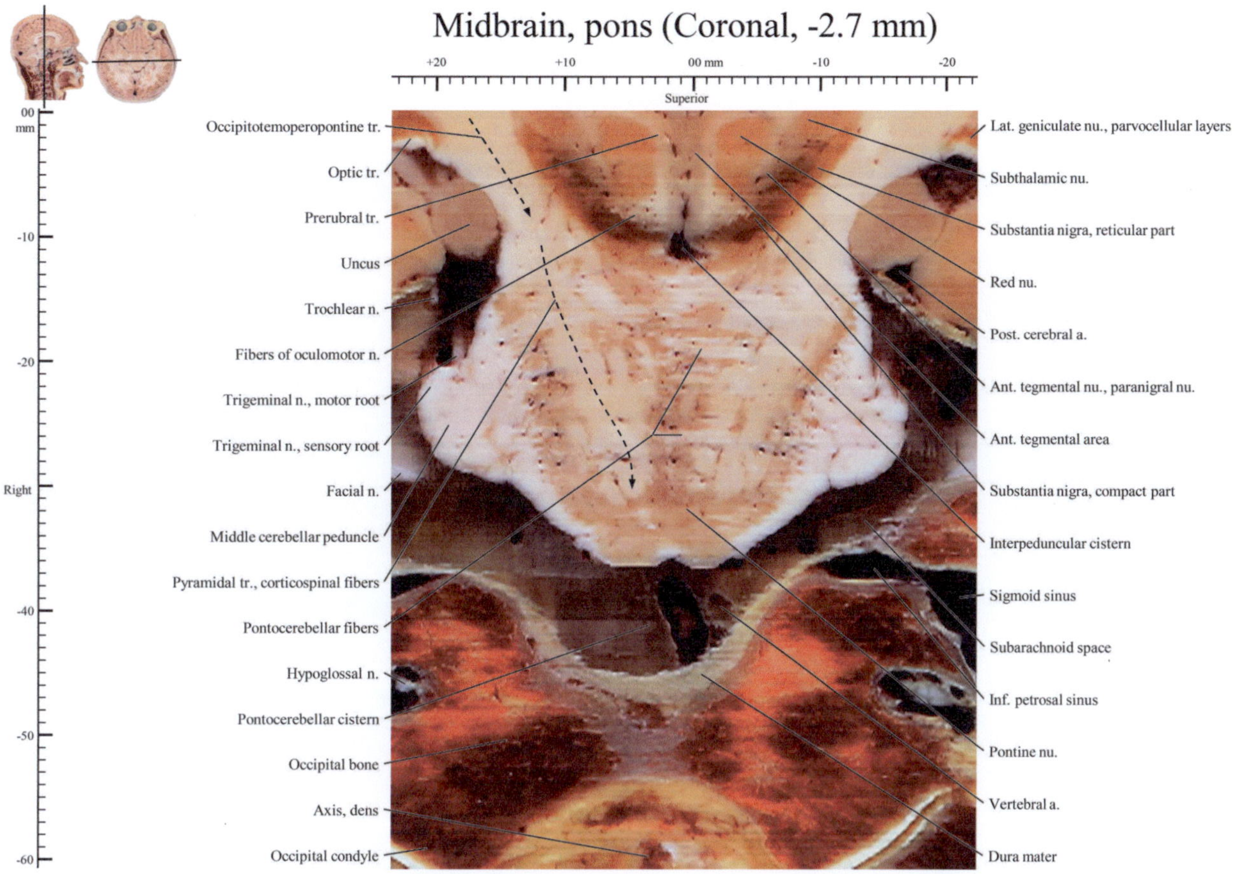

Midbrain, pons (Coronal, -3.6 mm)

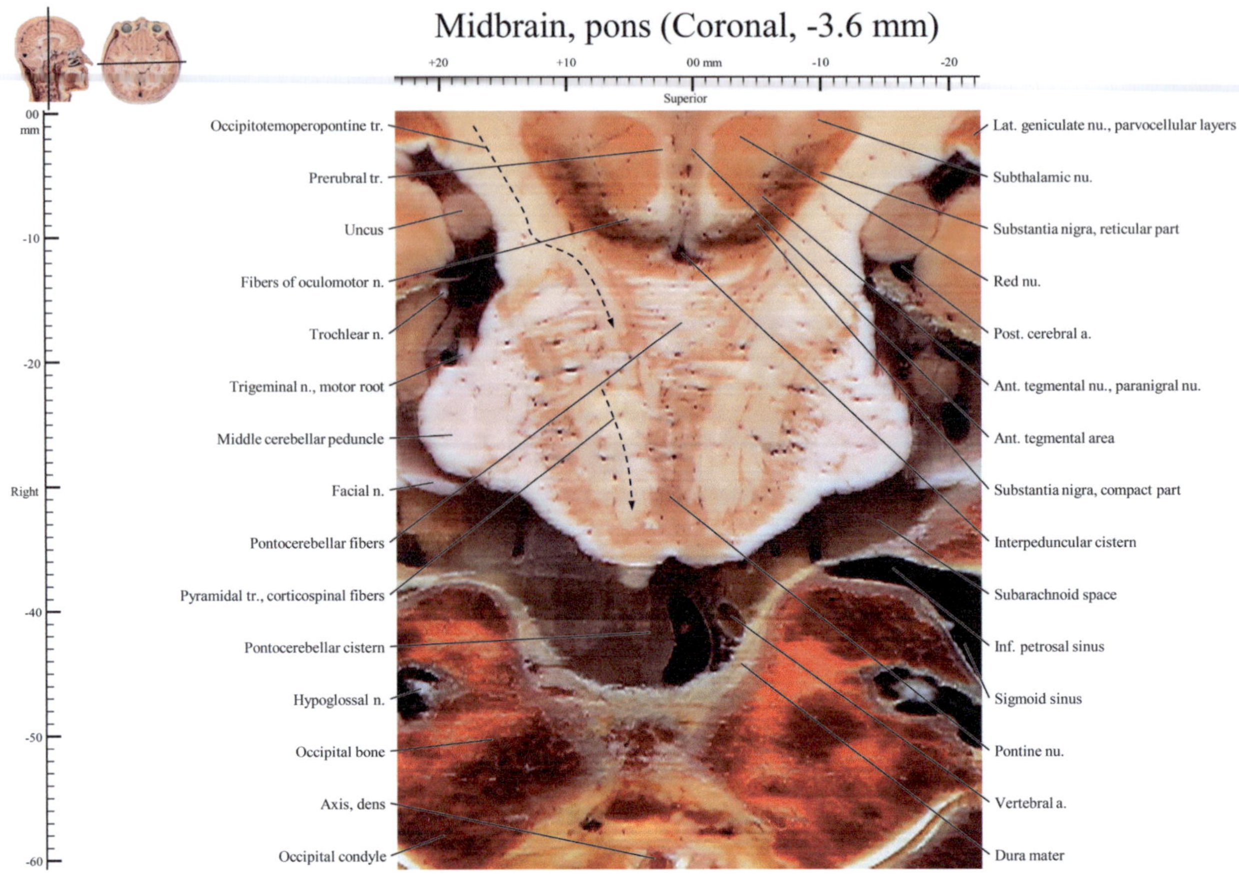

Midbrain, pons, medulla oblongata (Coronal, -4.5 mm)

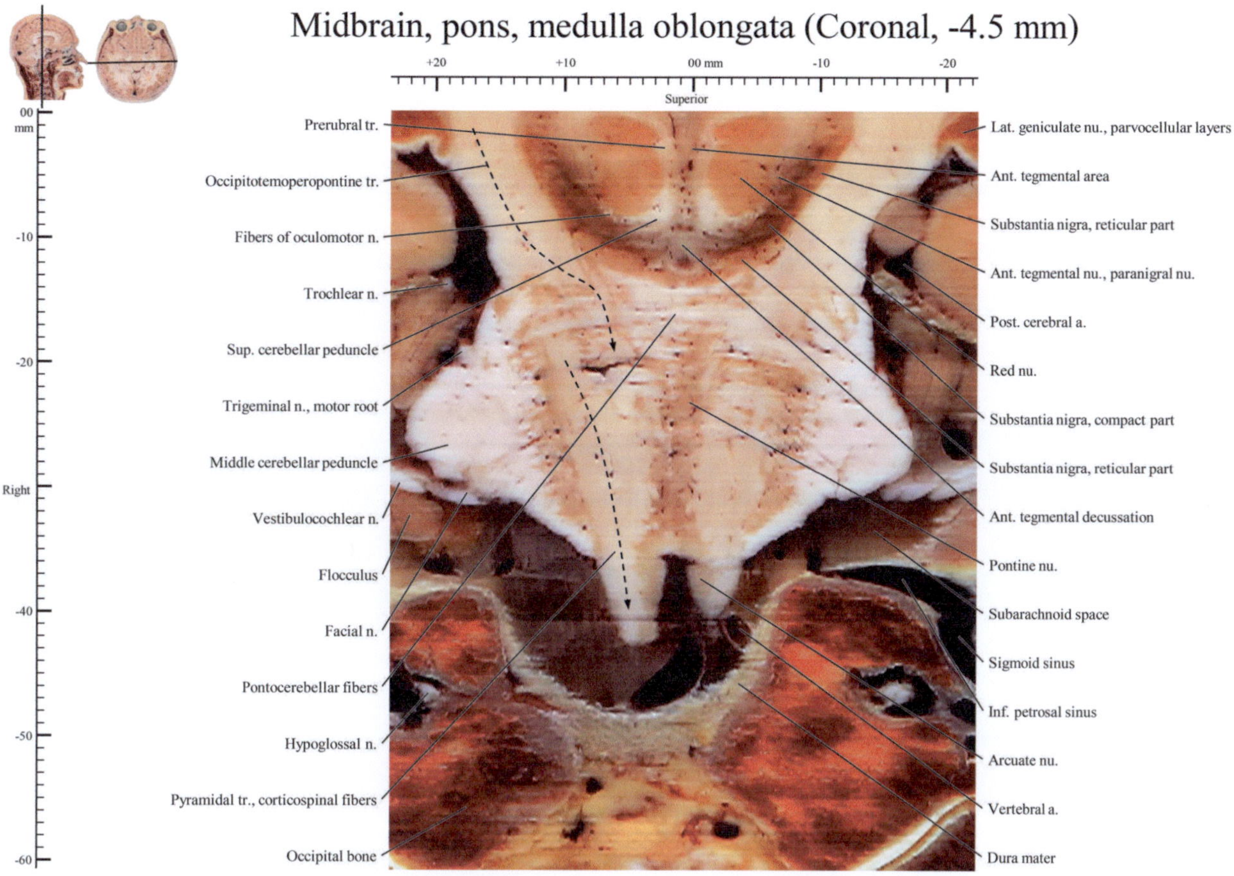

Midbrain, pons, medulla oblongata (Coronal, -5.4 mm)

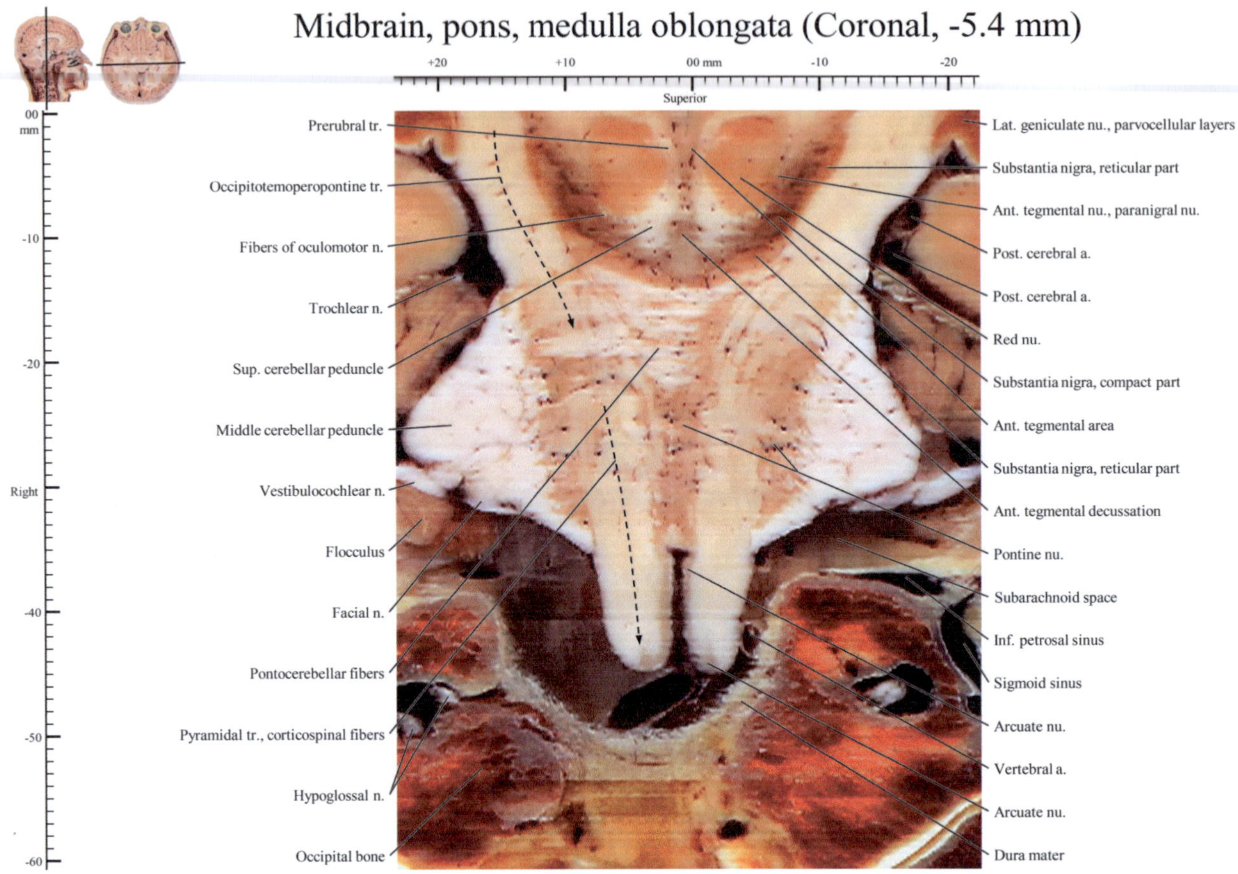

Midbrain, pons, medulla oblongata (Coronal, -6.3 mm)

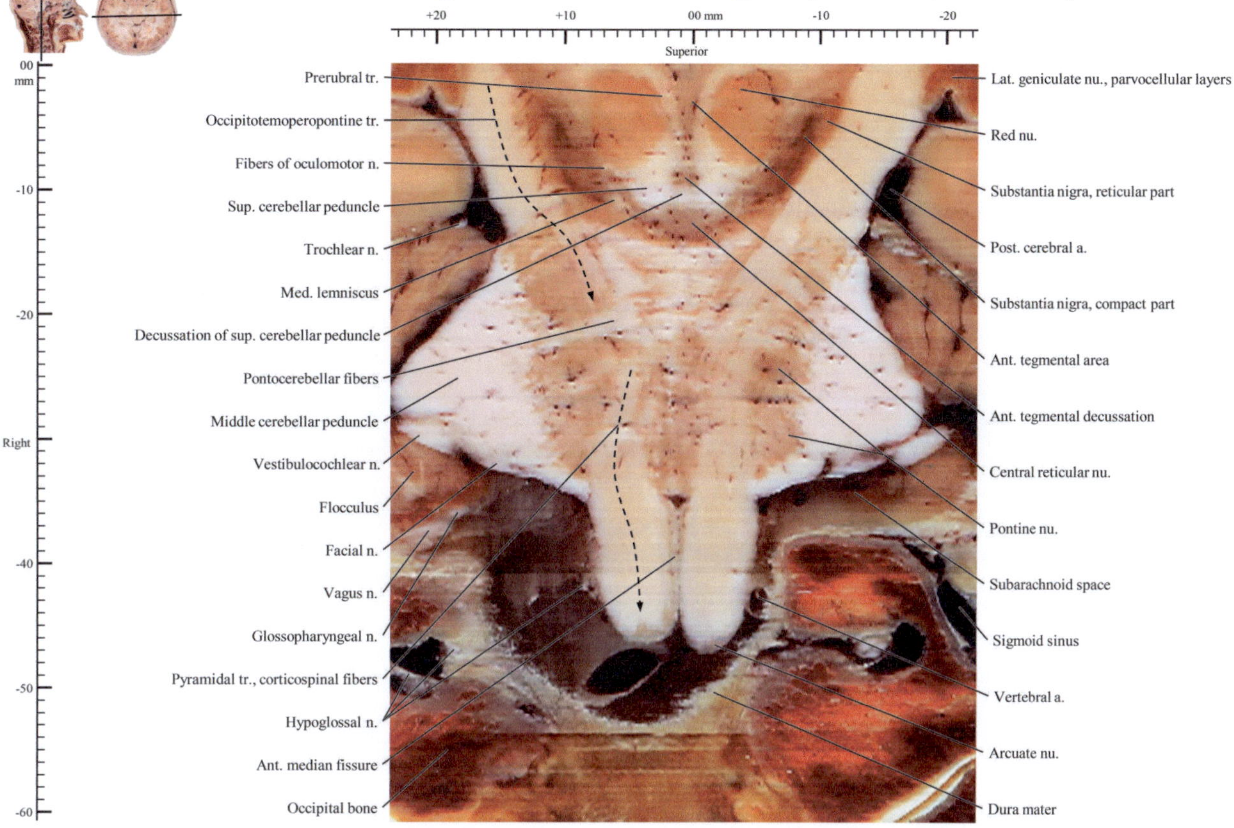

Midbrain, pons, medulla oblongata (Coronal, -7.2 mm)

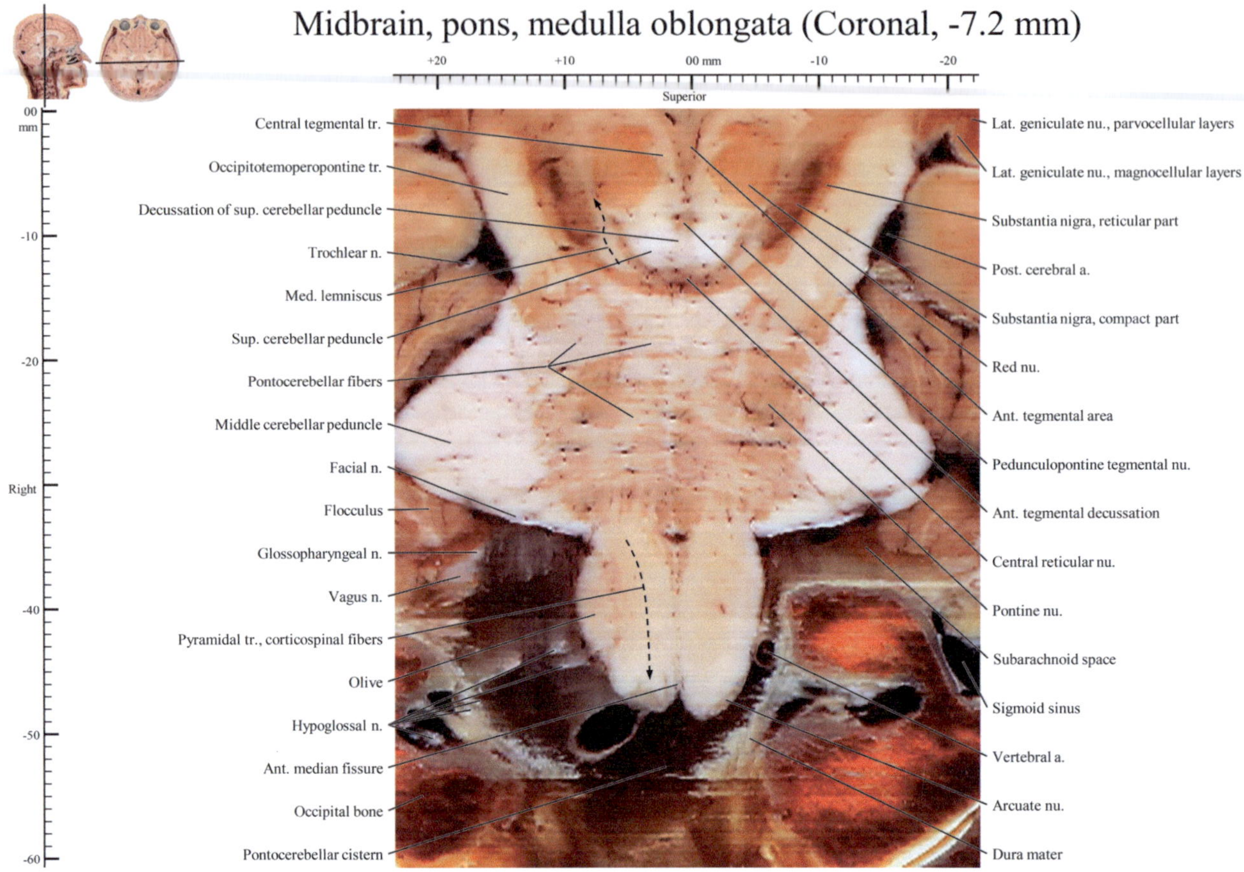

Midbrain, pons, medulla oblongata (Coronal, -8.1 mm)

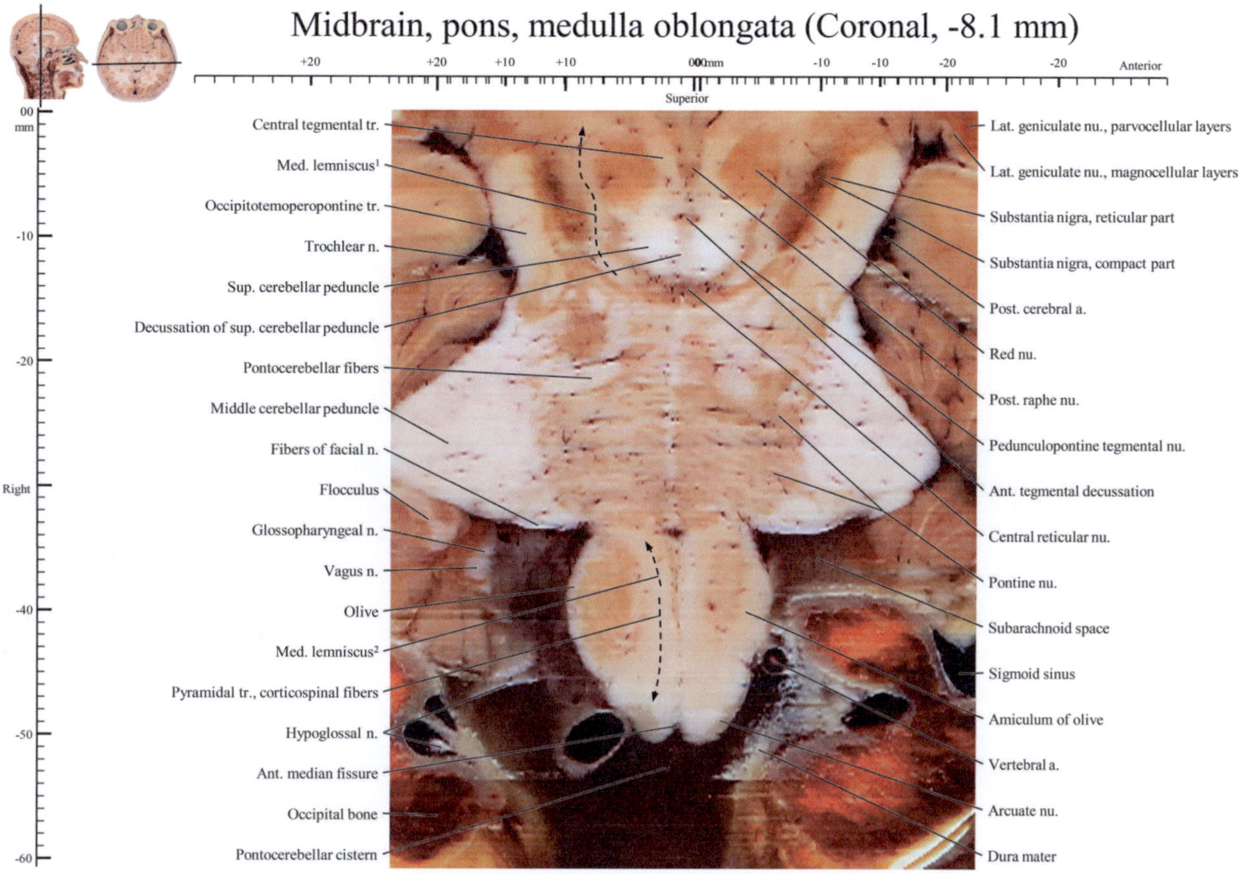

Midbrain, pons, medulla oblongata (Coronal, -9.0 mm)

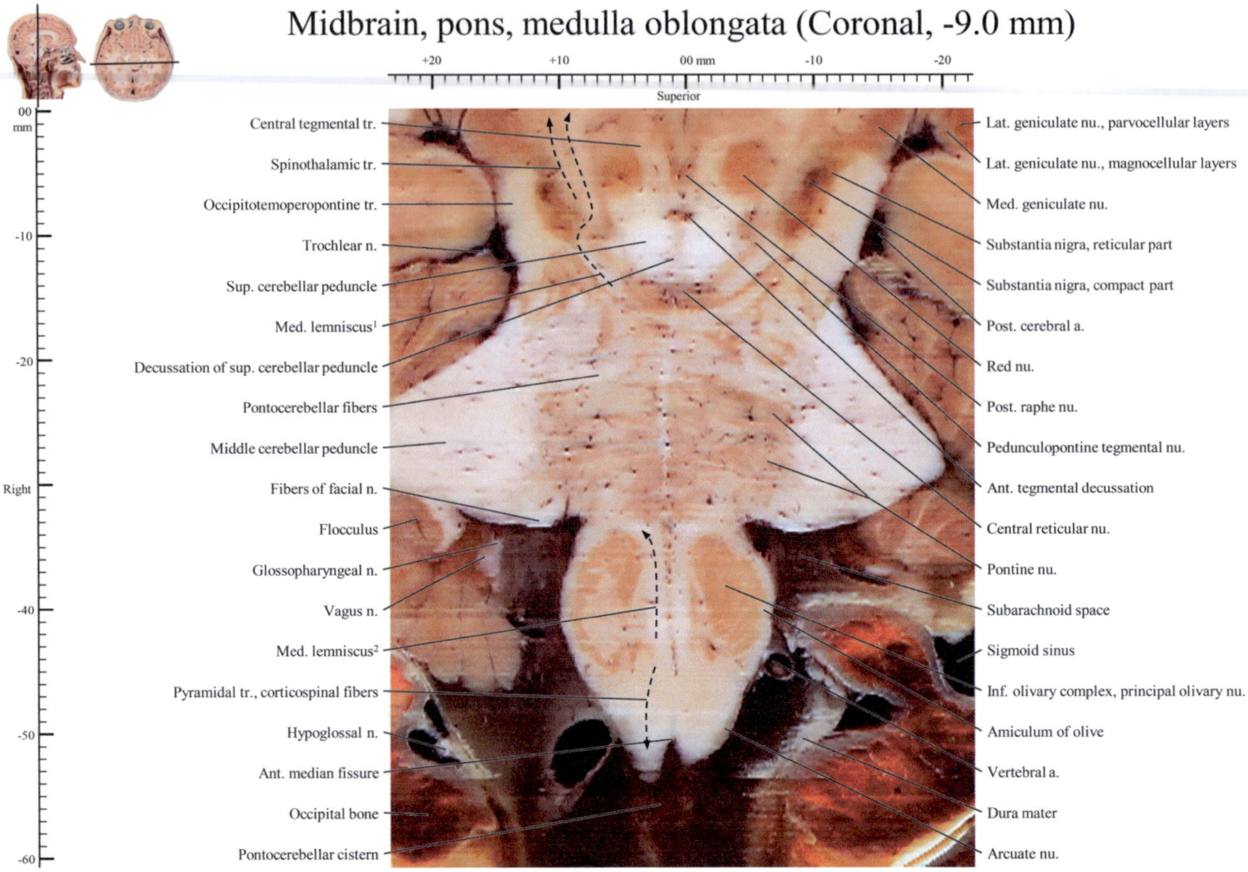

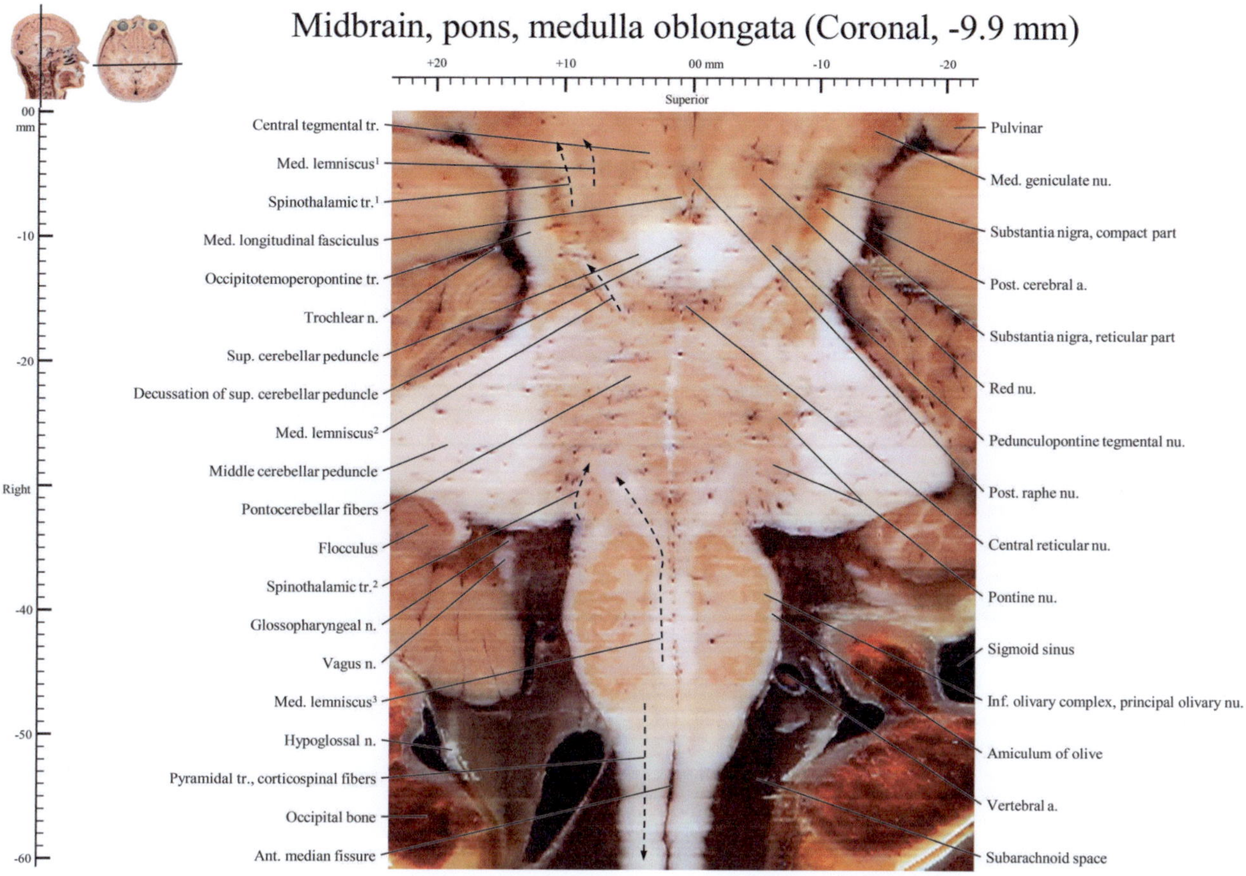
Midbrain, pons, medulla oblongata (Coronal, -9.9 mm)
+20
+10
00 mm
-10
-20
Superior
00 mm
-10
-20
Right
-40
-50
-60
Central tegmental tr.
Med. lemniscus[1]
Spinothalamic tr.[1]
Med. longitudinal fasciculus
Occipitotemoperopontine tr.
Trochlear n.
Sup. cerebellar peduncle
Decussation of sup. cerebellar peduncle
Med. lemniscus[2]
Middle cerebellar peduncle
Pontocerebellar fibers
Flocculus
Spinothalamic tr.[2]
Glossopharyngeal n.
Vagus n.
Med. lemniscus[3]
Hypoglossal n.
Pyramidal tr., corticospinal fibers
Occipital bone
Ant. median fissure
Pulvinar
Med. geniculate nu.
Substantia nigra, compact part
Post. cerebral a.
Substantia nigra, reticular part
Red nu.
Pedunculopontine tegmental nu.
Post. raphe nu.
Central reticular nu.
Pontine nu.
Sigmoid sinus
Inf. olivary complex, principal olivary nu.
Amiculum of olive
Vertebral a.
Subarachnoid space

Midbrain, pons, medulla oblongata (Coronal, -10.8 mm)

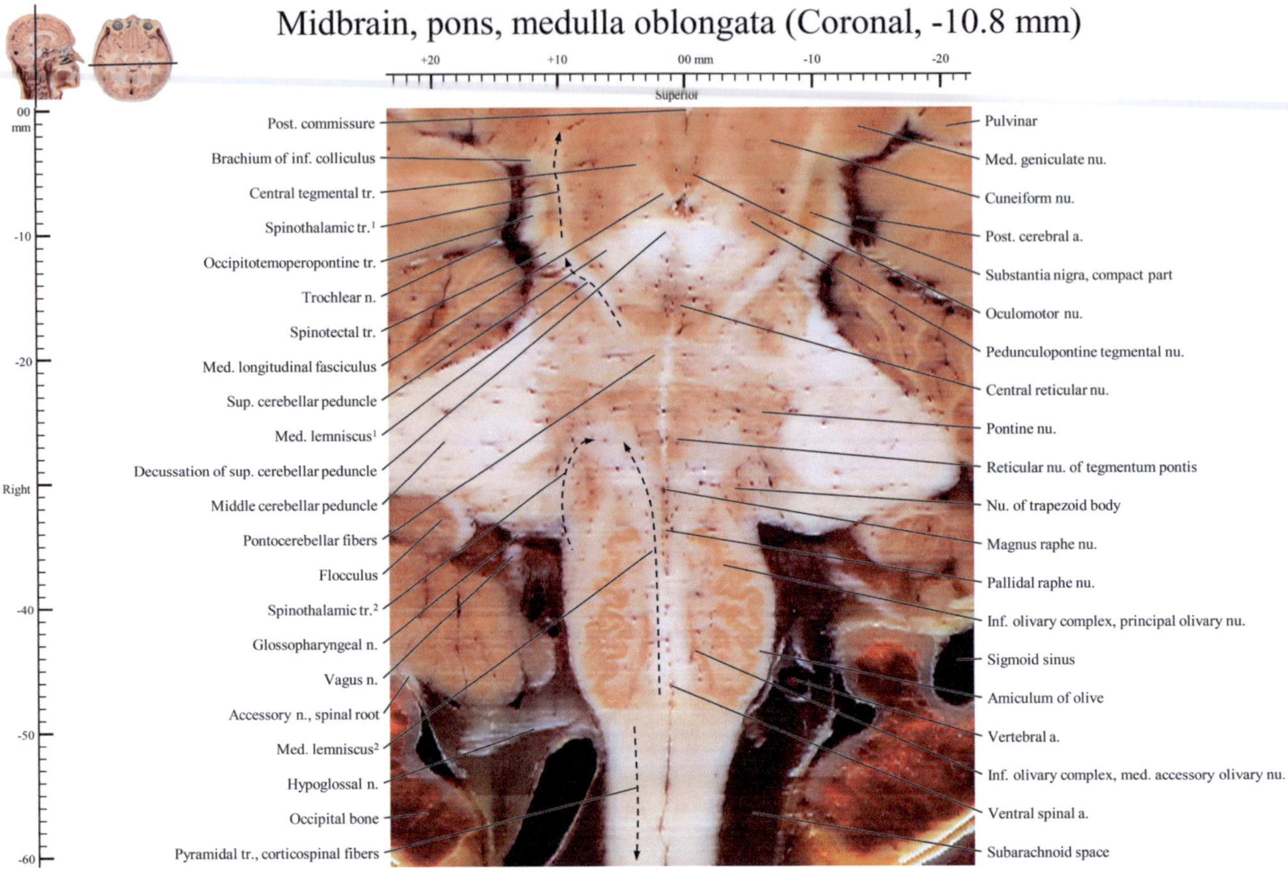

Midbrain, pons, medulla oblongata (Coronal, -11.7 mm)

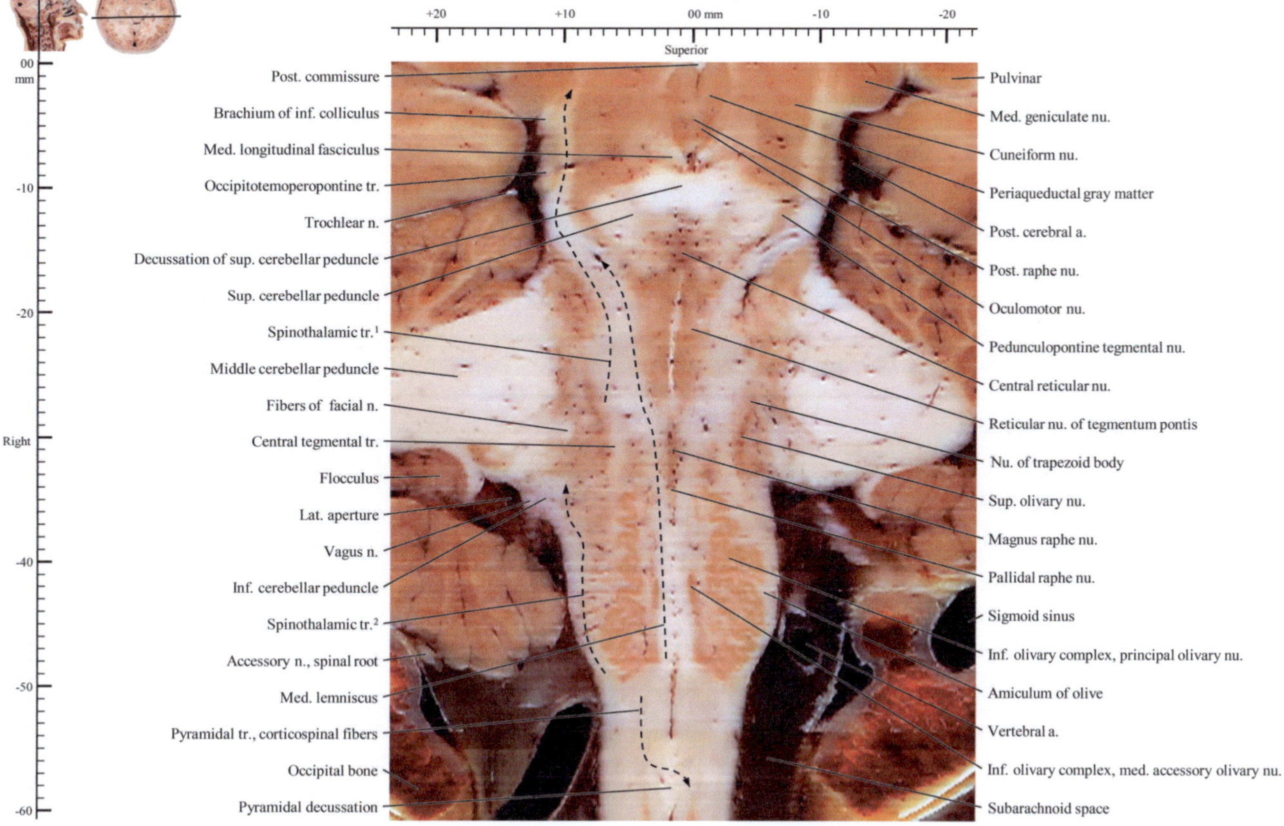

Midbrain, pons, medulla oblongata (Coronal, -12.6 mm)

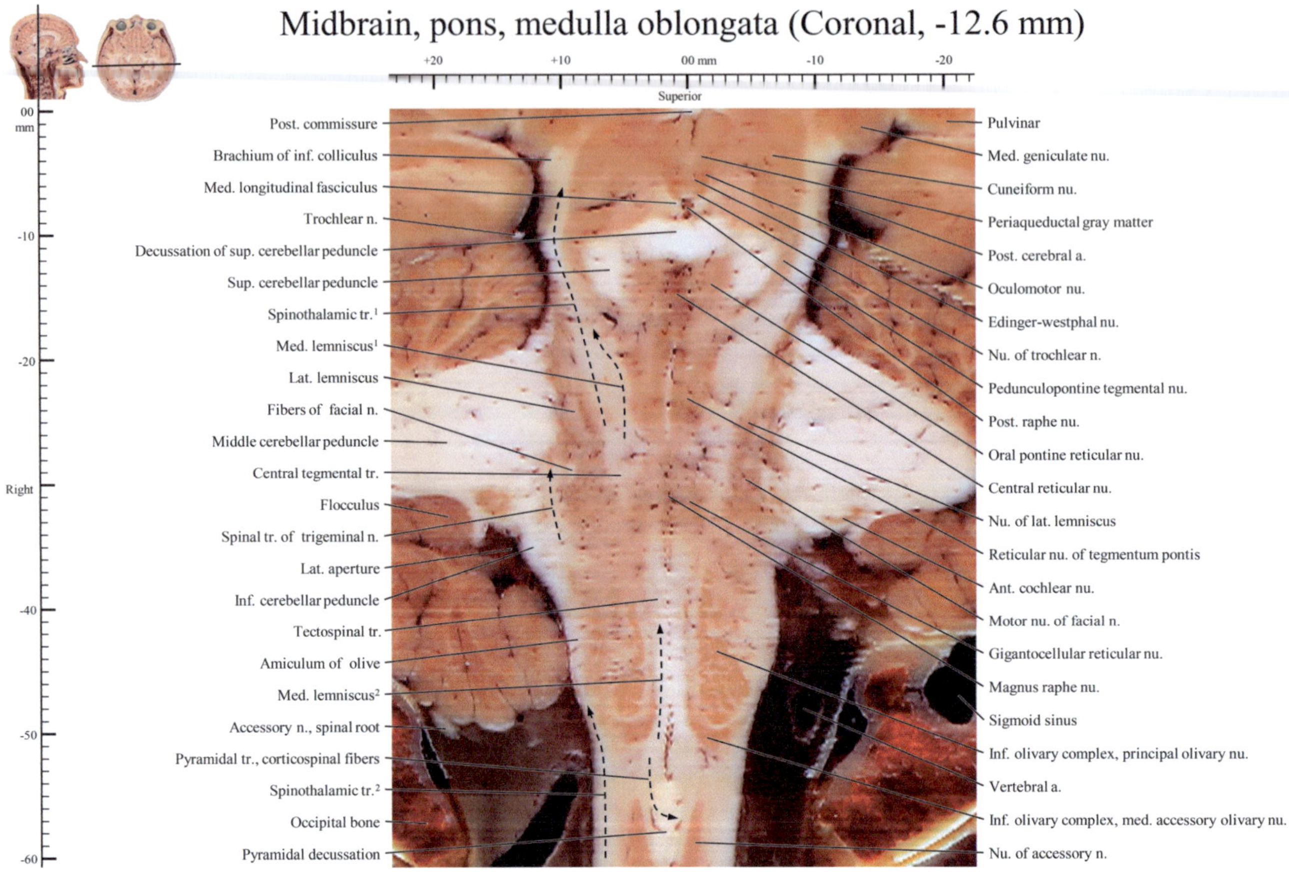

Midbrain, pons, medulla oblongata (Coronal, -13.5 mm)

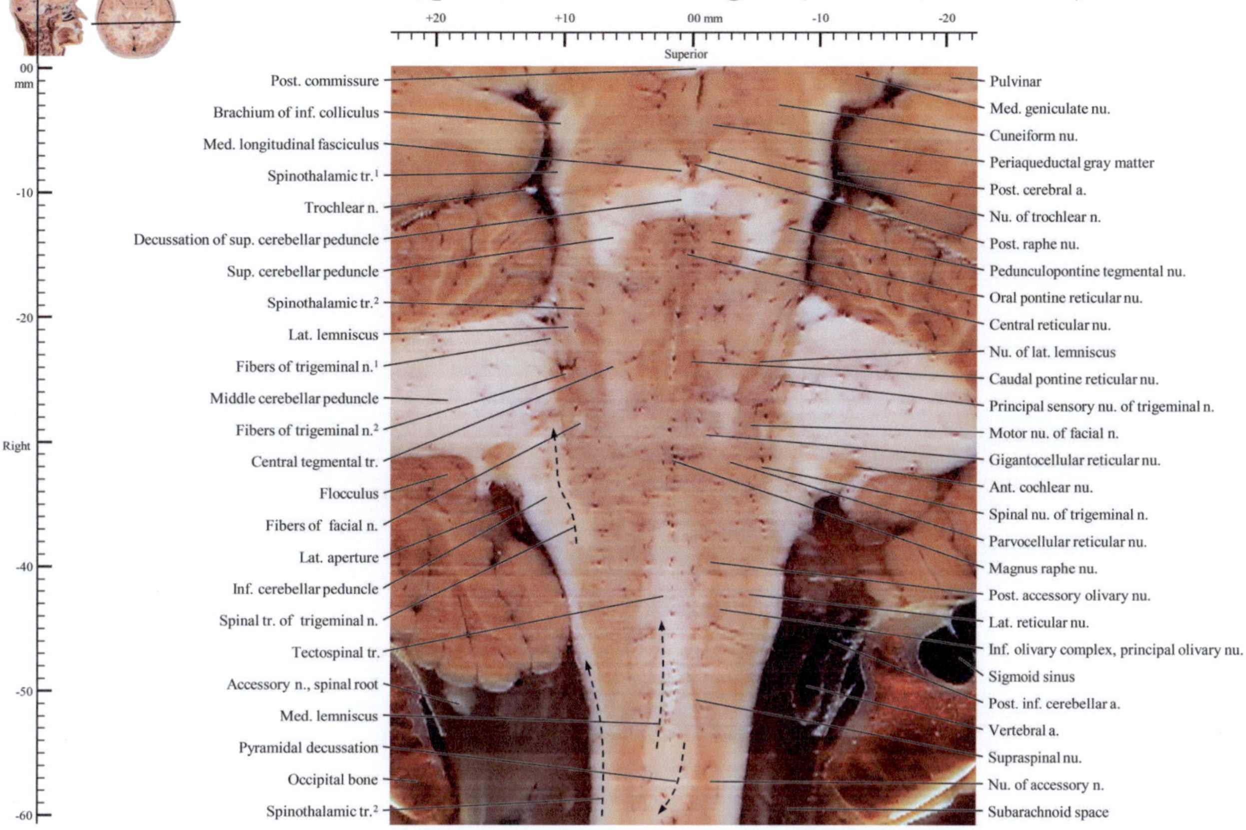

Midbrain, pons, medulla oblongata (Coronal, -14.4 mm)

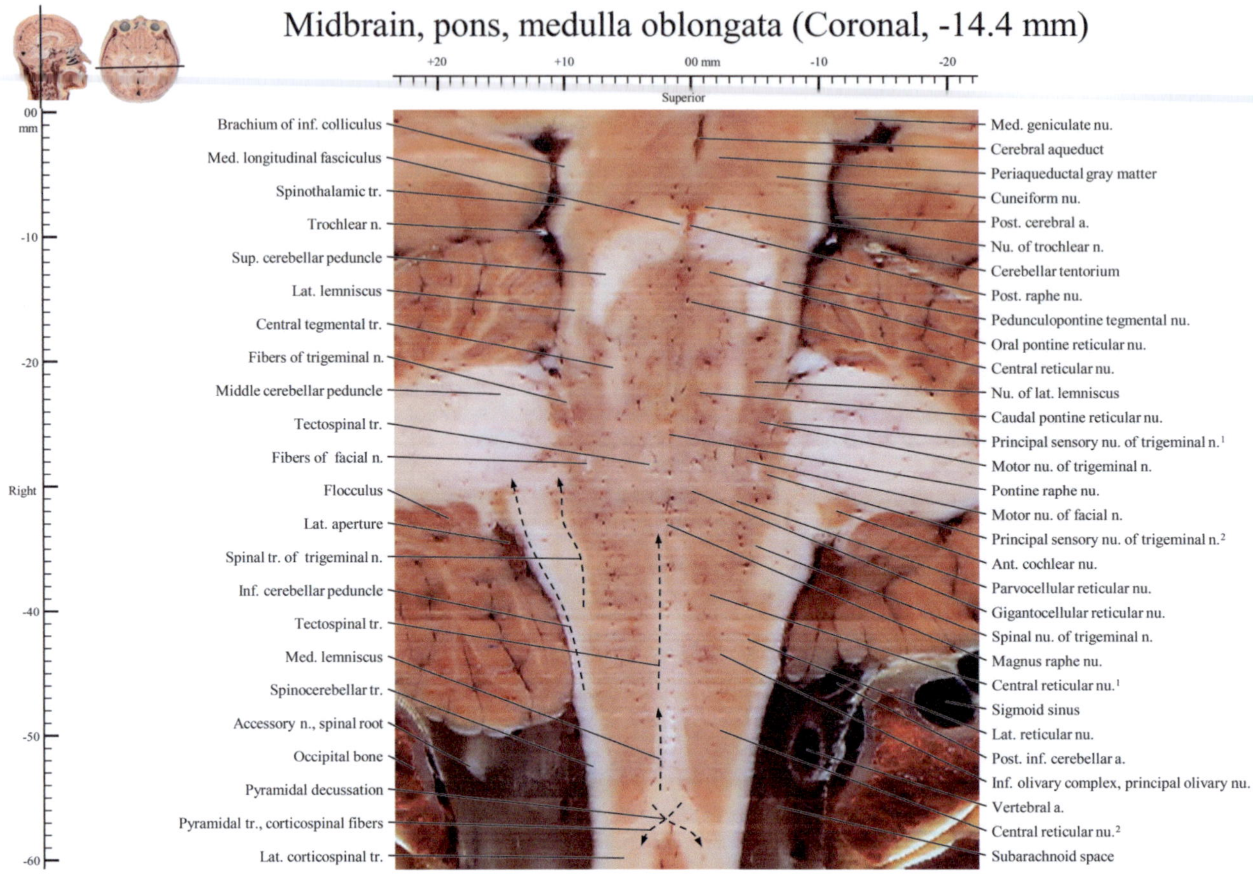

Midbrain, pons, medulla oblongata (Coronal, -15.3 mm)

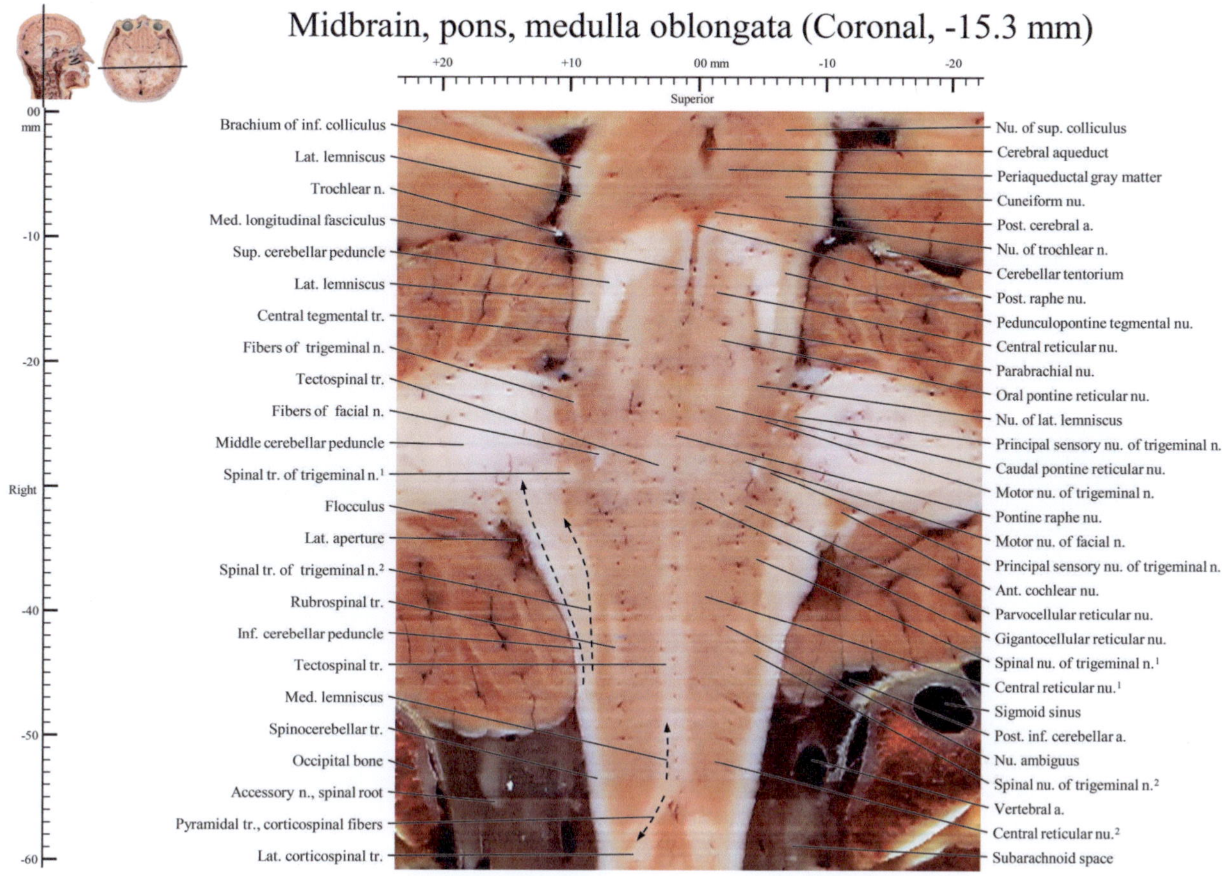

Midbrain, pons, medulla oblongata (Coronal, -16.2 mm)

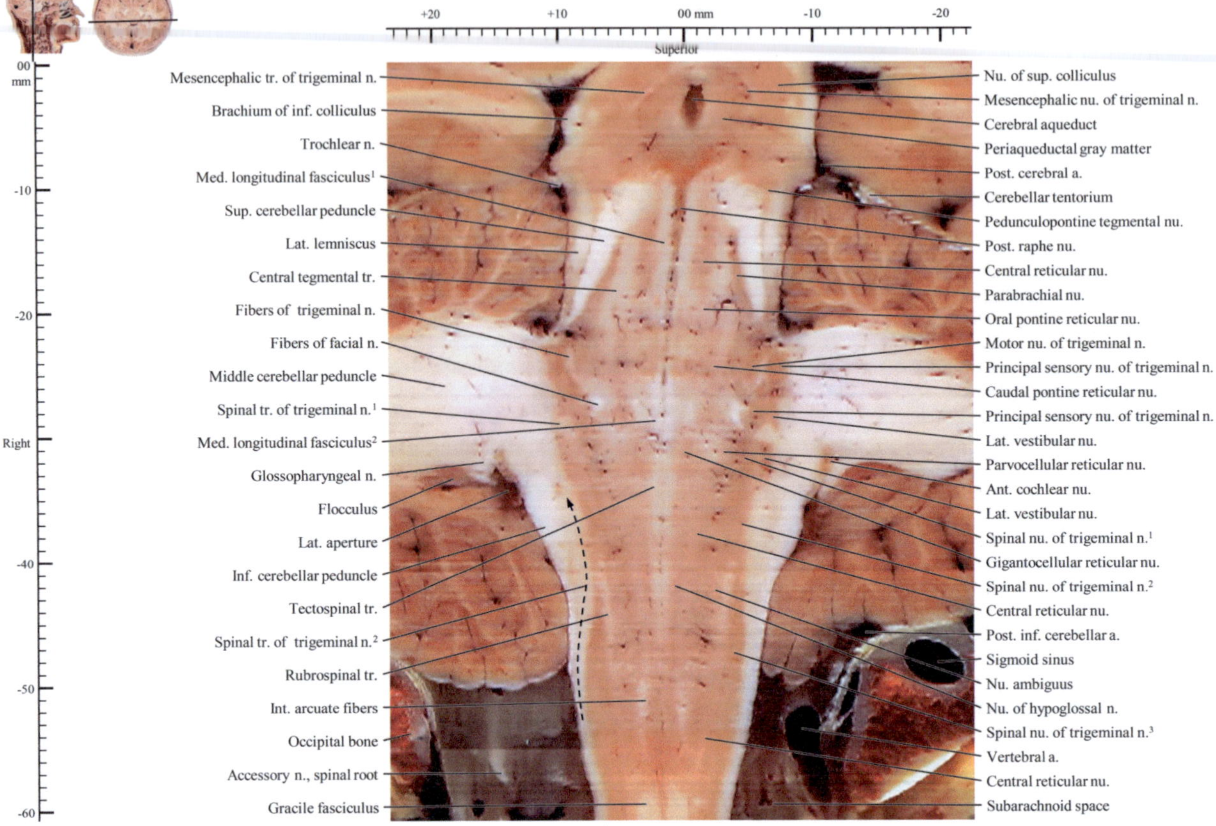

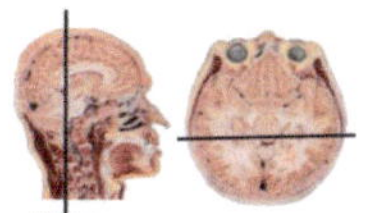

Midbrain, pons, medulla oblongata (Coronal, -17.1 mm)

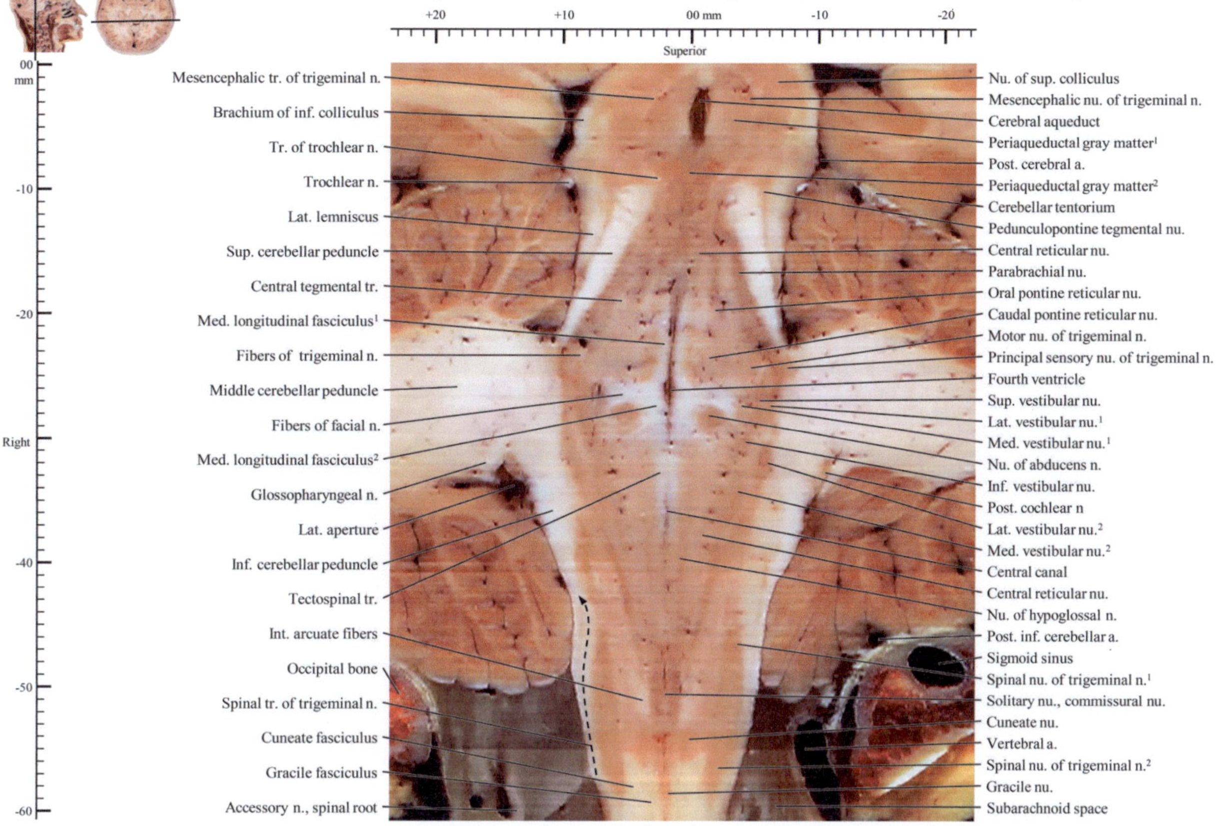

Midbrain, pons, medulla oblongata (Coronal, -18.0 mm)

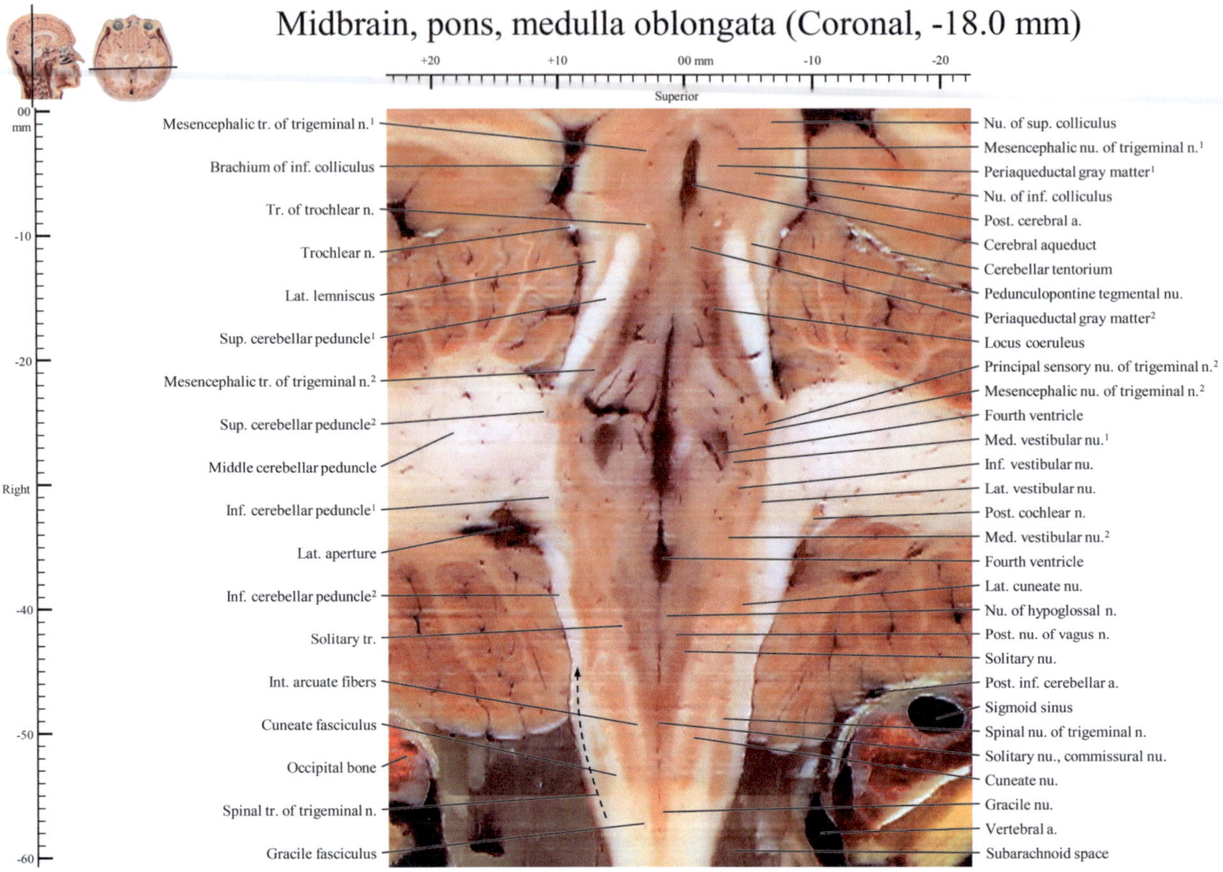

Midbrain, medulla oblongata (Coronal, -18.9 mm)

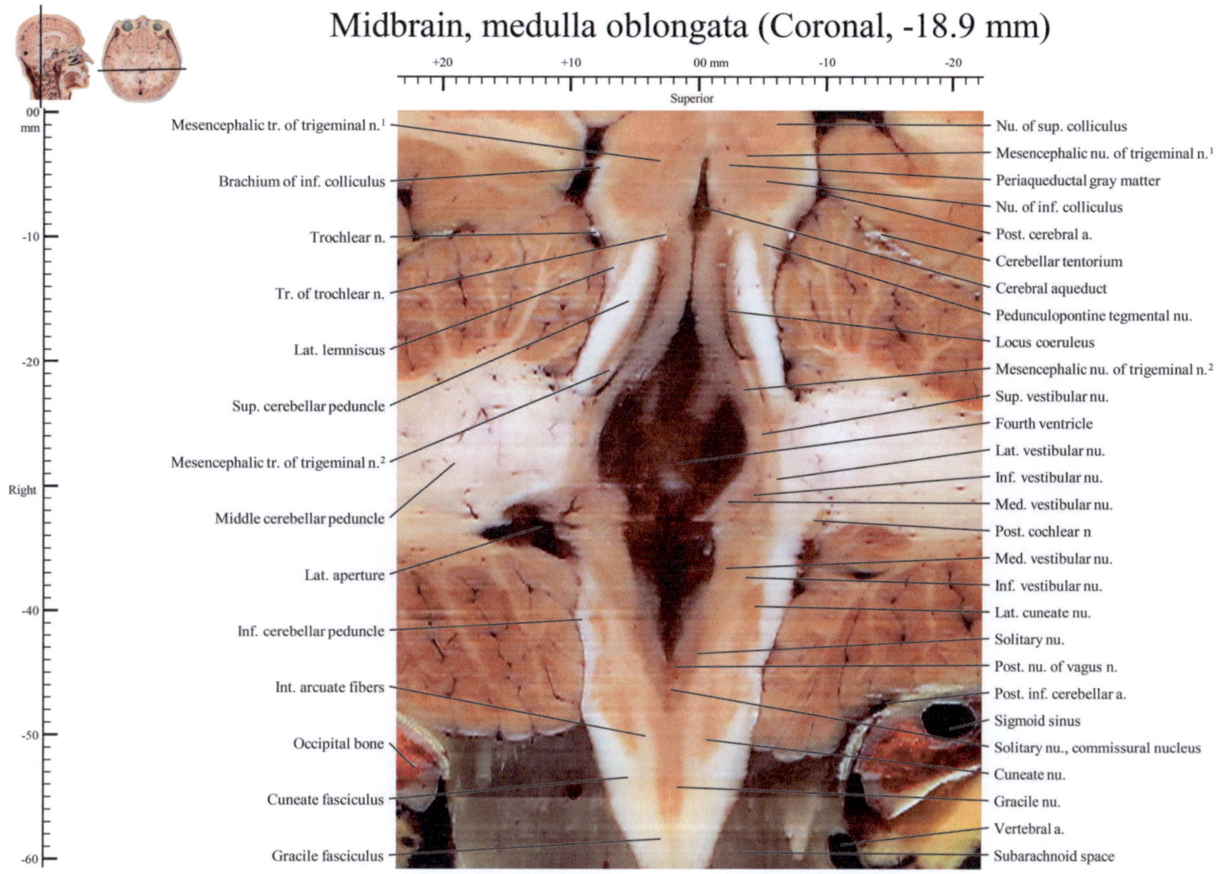

Midbrain, medulla oblongata (Coronal, -19.8 mm)

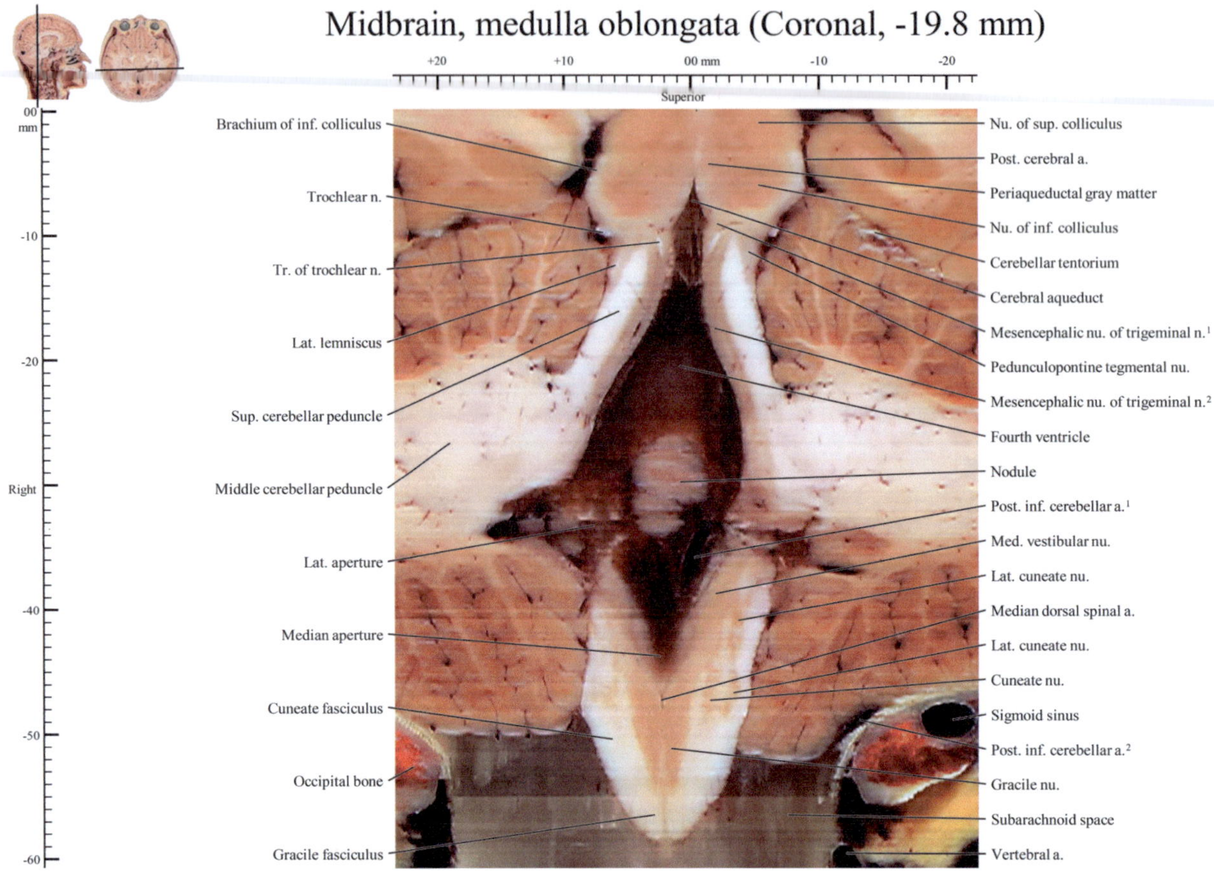

Midbrain, medulla oblongata (Coronal, -20.7 mm)

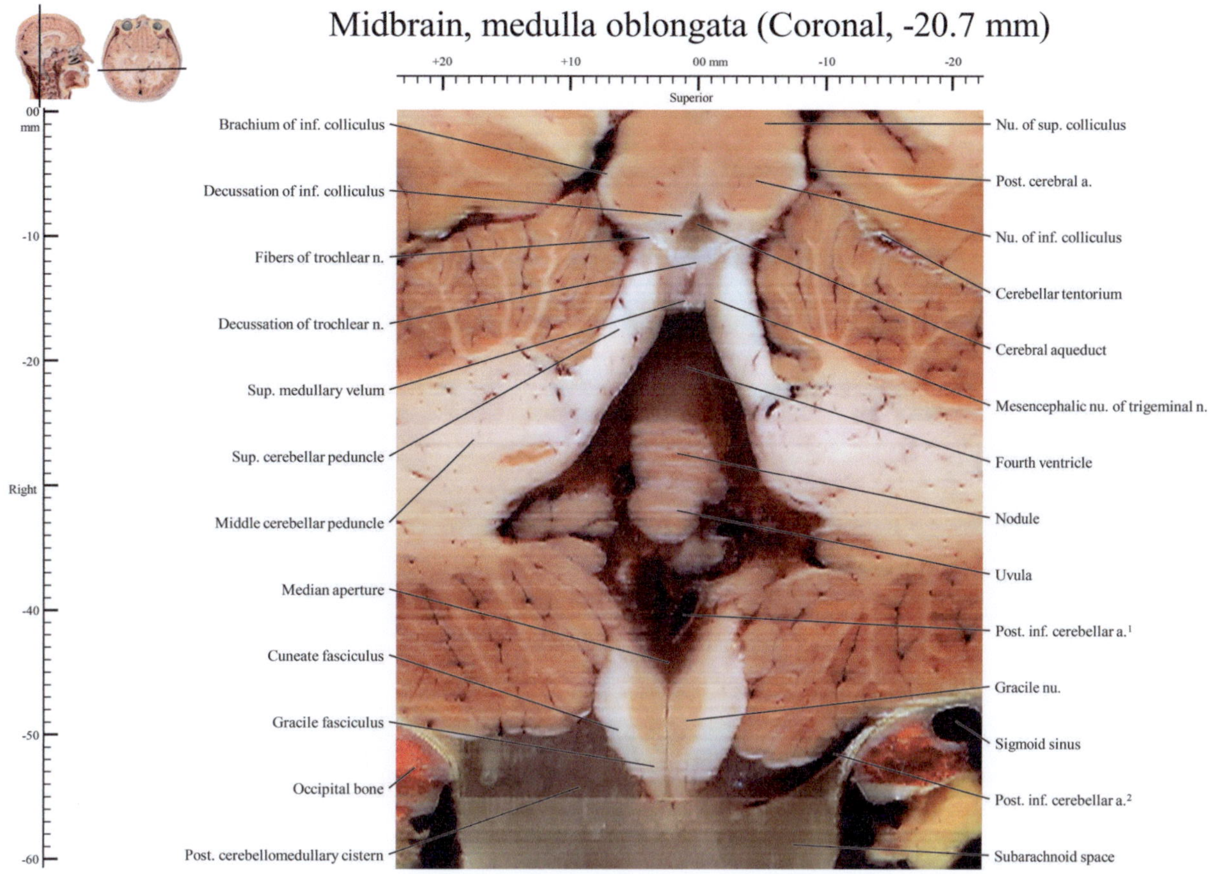

Midbrain, medulla oblongata (Coronal, -21.6 mm)

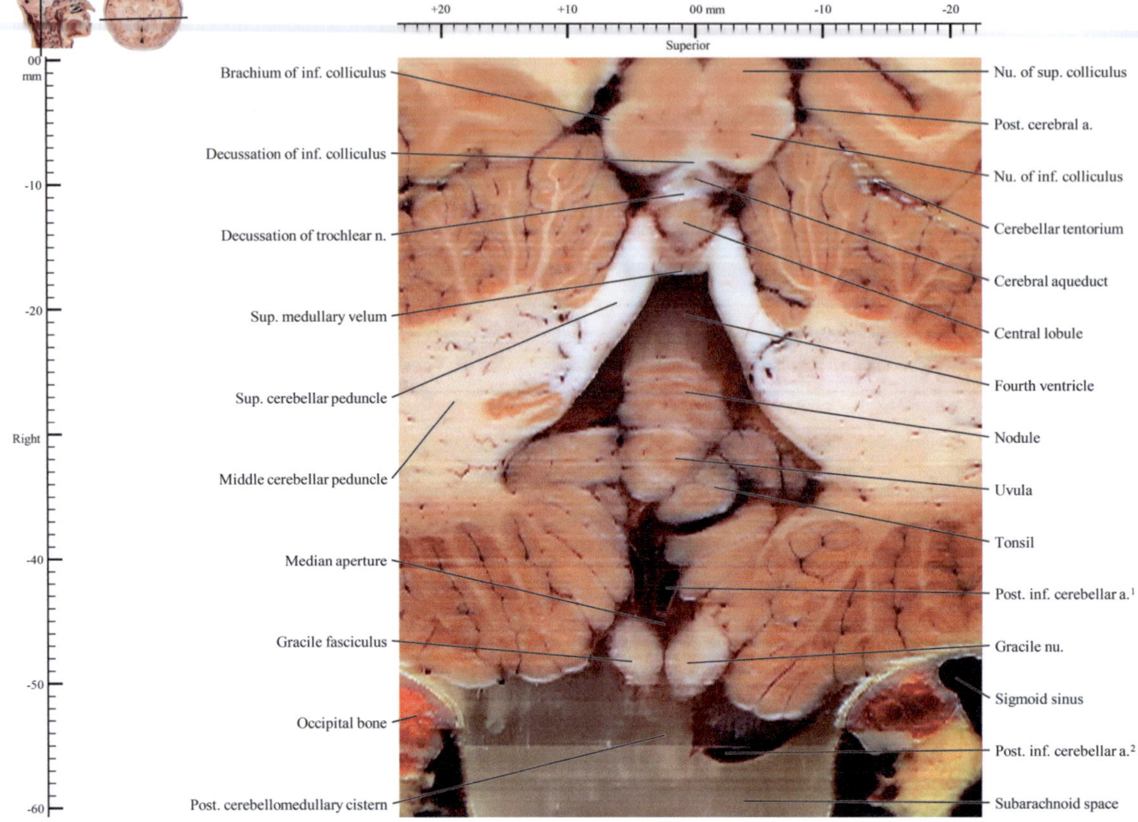

Midbrain (Coronal, -22.5 mm)

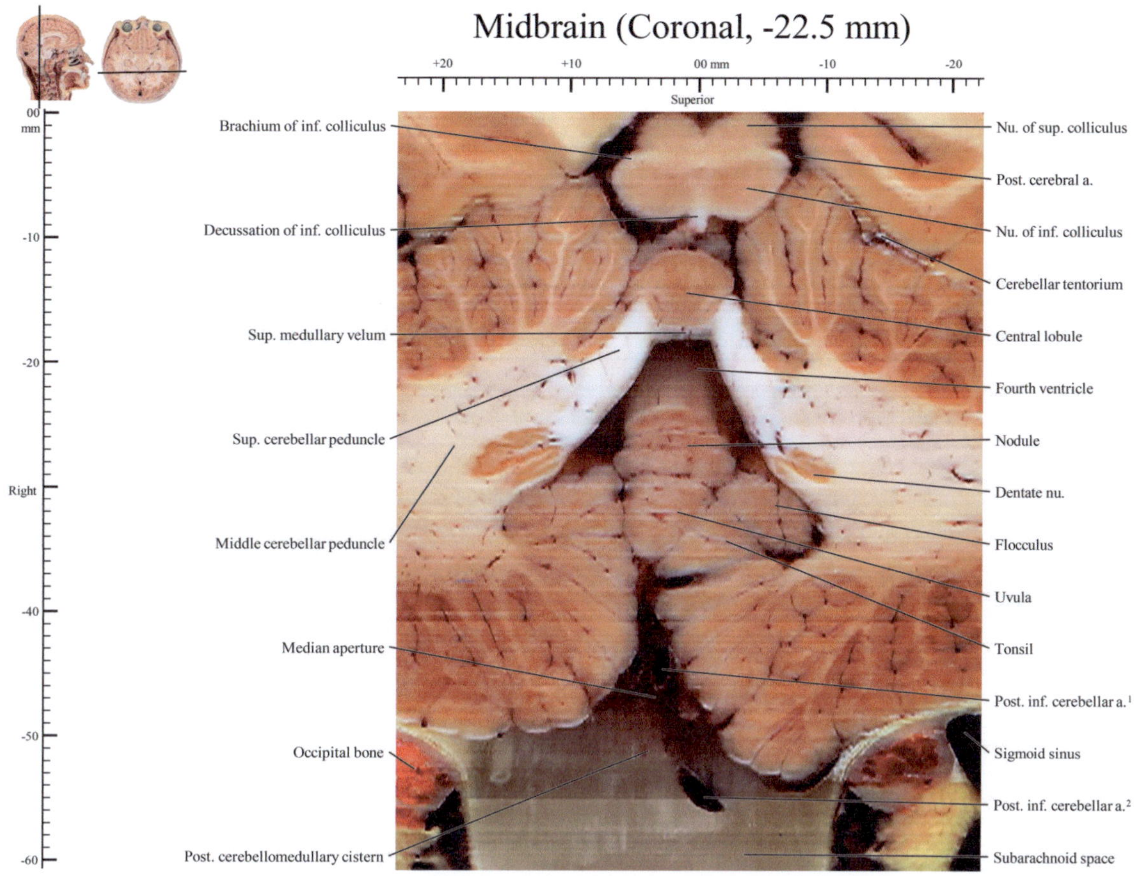

4 Sagittal Brainstem (Left Side)

Images of 0.0 mm ~ −9.9 mm, Midbrain, pons, medulla oblongata

Images of −10.8 mm ~ 11.7 mm, Midbrain, medulla oblongata

Images of −18.0 mm, Pons

All images, 0.06 mm × 0.06 mm sized-pixel and 0.9 mm intervals

Abbreviation: a., artery; ant., anterior; inf., inferior; int., internal; lat., lateral; m., muscle; med., medial; n., nerve; nu., nucleus; post., posterior; sup., superior; tr., tract

J. S. Park, Y. You, *Cross-Sectional Atlas of Human Brainstem*, https://doi.org/10.1007/978-981-19-9519-4_4

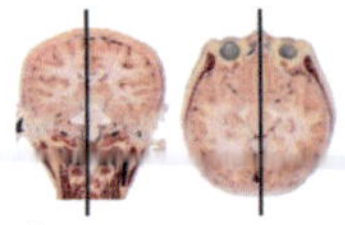

Midbrain, pons, medulla oblongata (Sagittal, 0.0 mm)

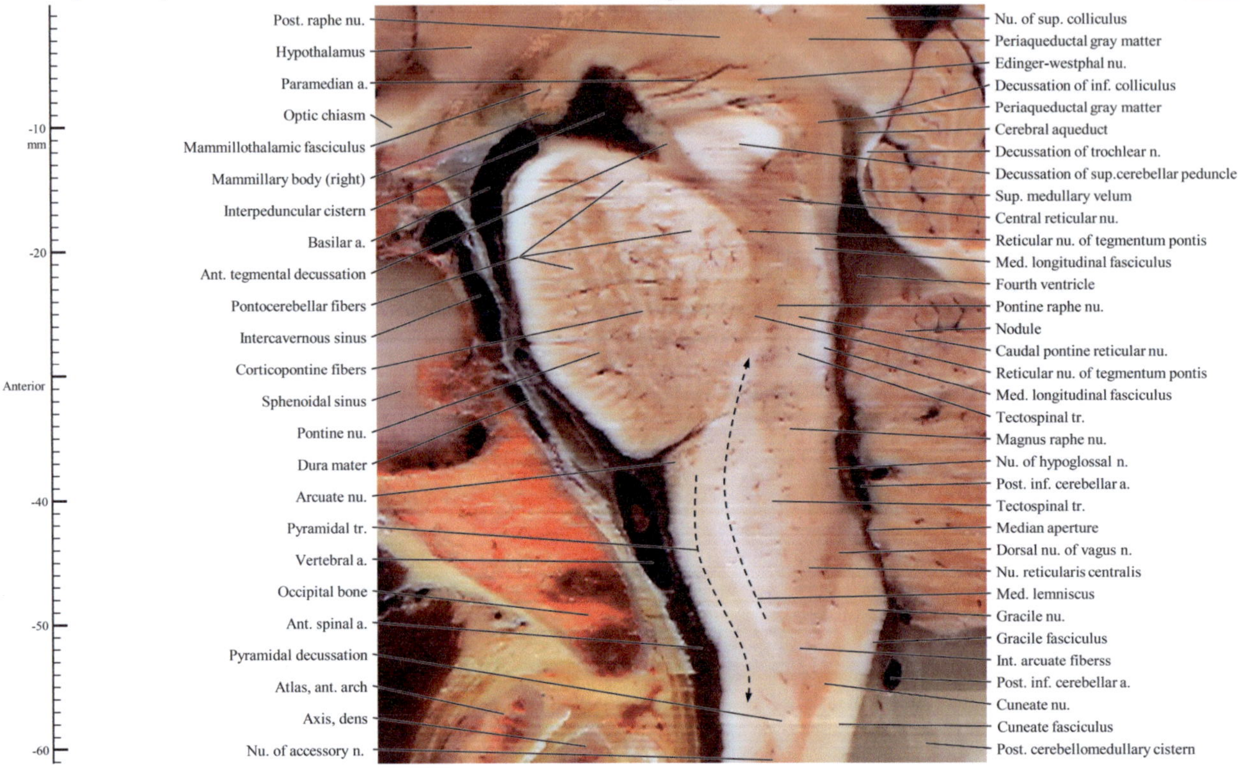

Midbrain, pons, medulla oblongata (Sagittal, -0.9 mm)

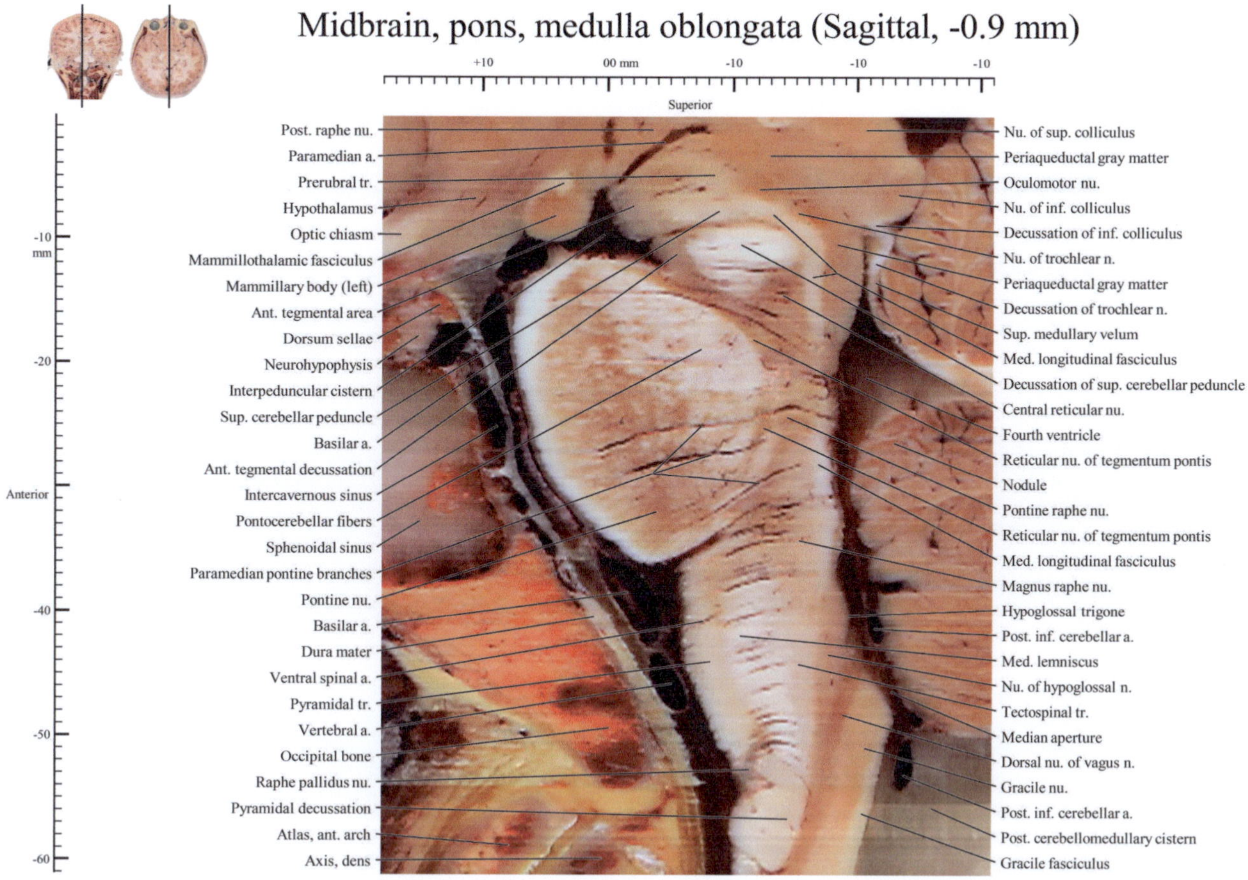

Midbrain, pons, medulla oblongata (Sagittal, -1.8 mm)

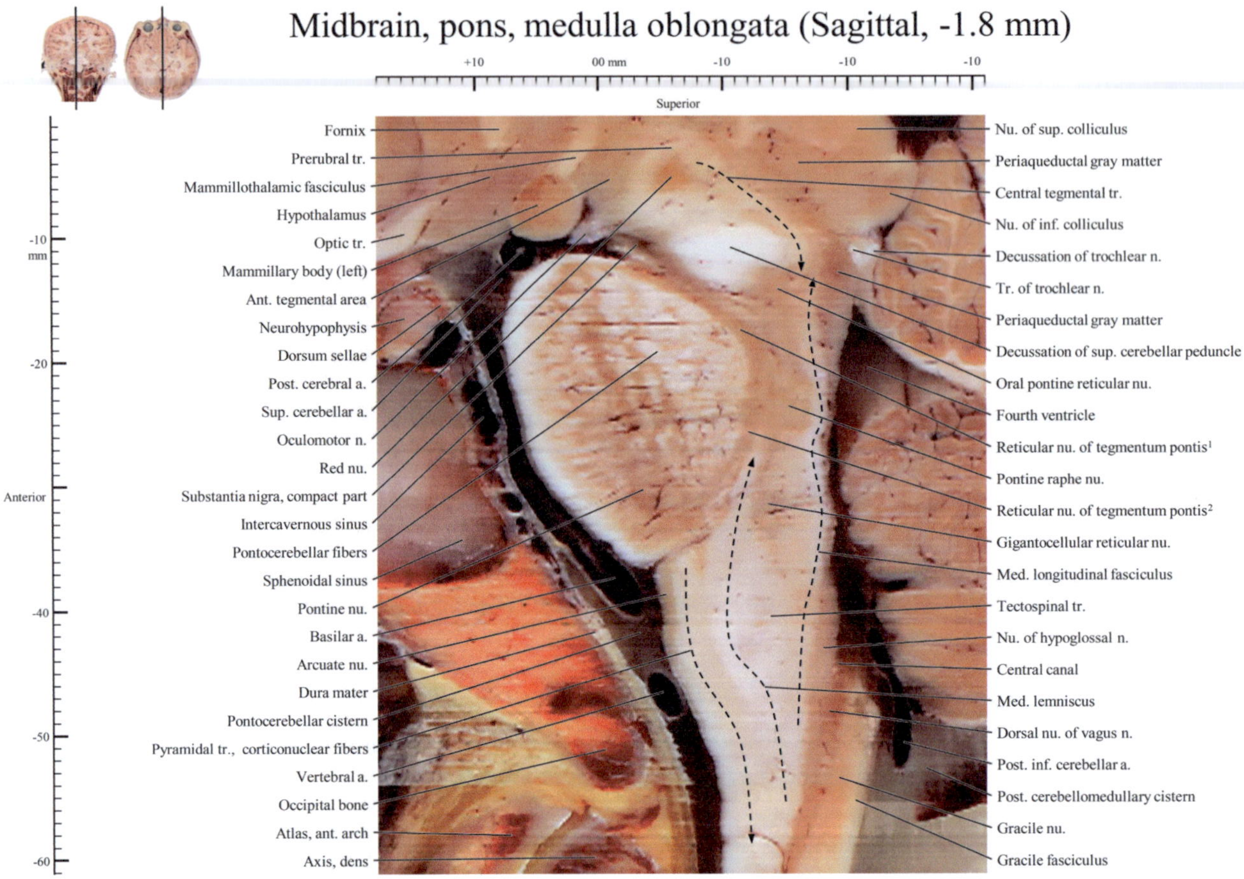

Midbrain, pons, medulla oblongata (Sagittal, -2.7 mm)

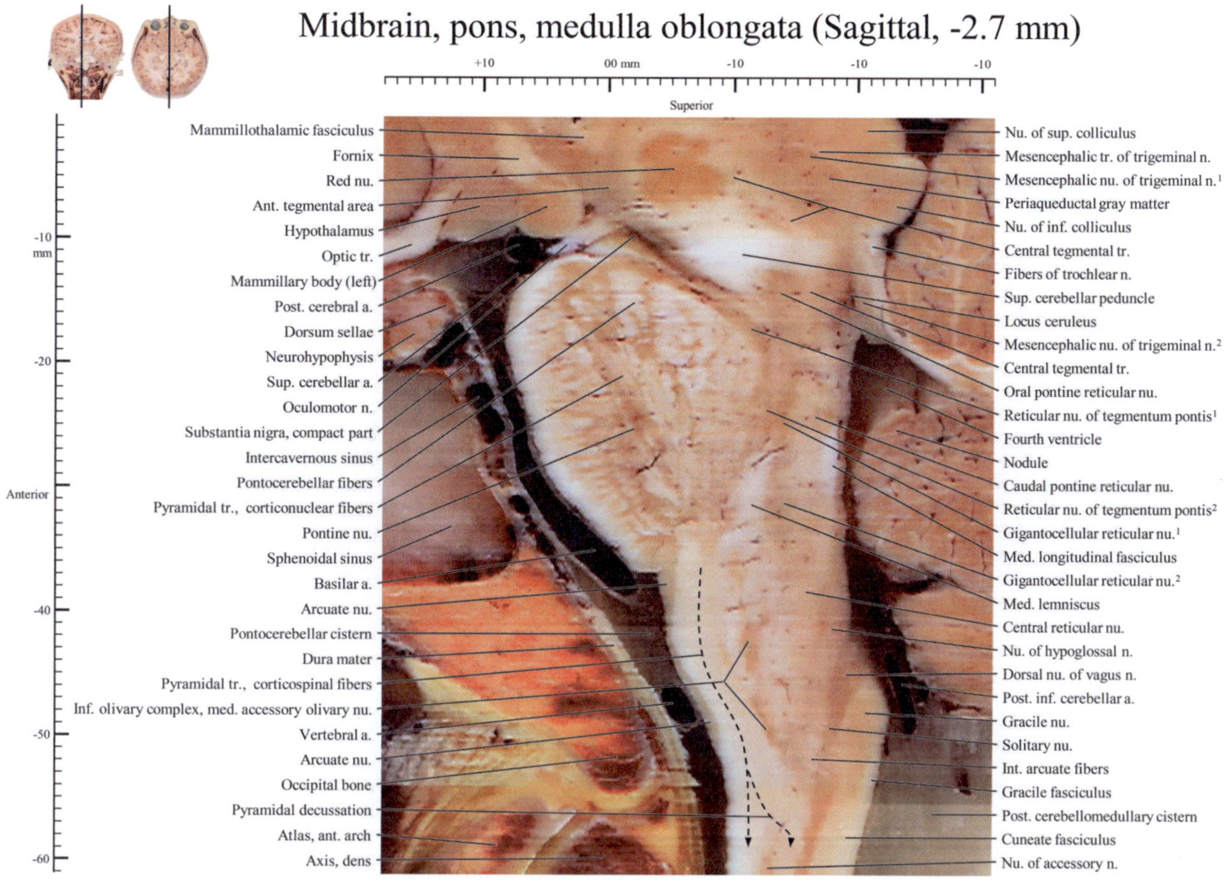

Midbrain, pons, medulla oblongata (Sagittal, -3.6 mm)

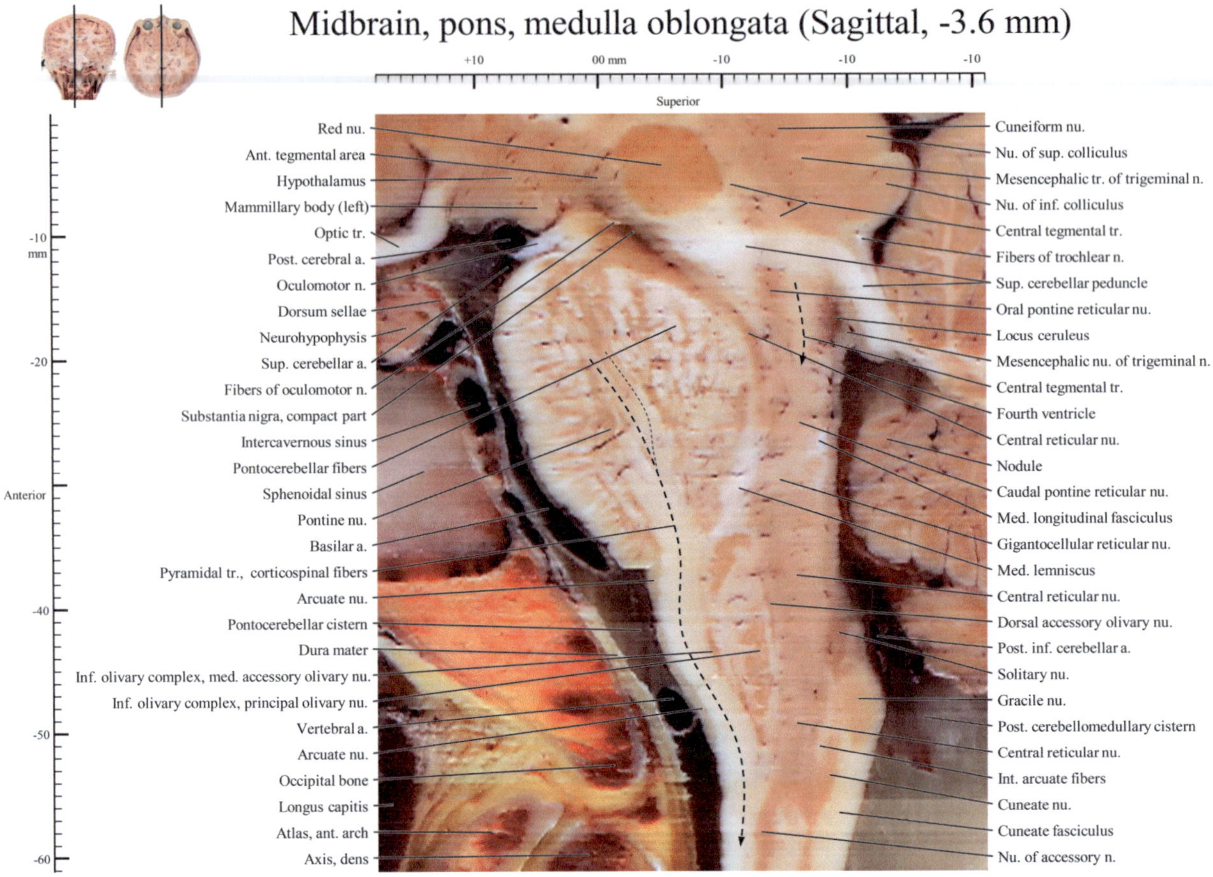

Midbrain, pons, medulla oblongata (Sagittal, -4.5 mm)

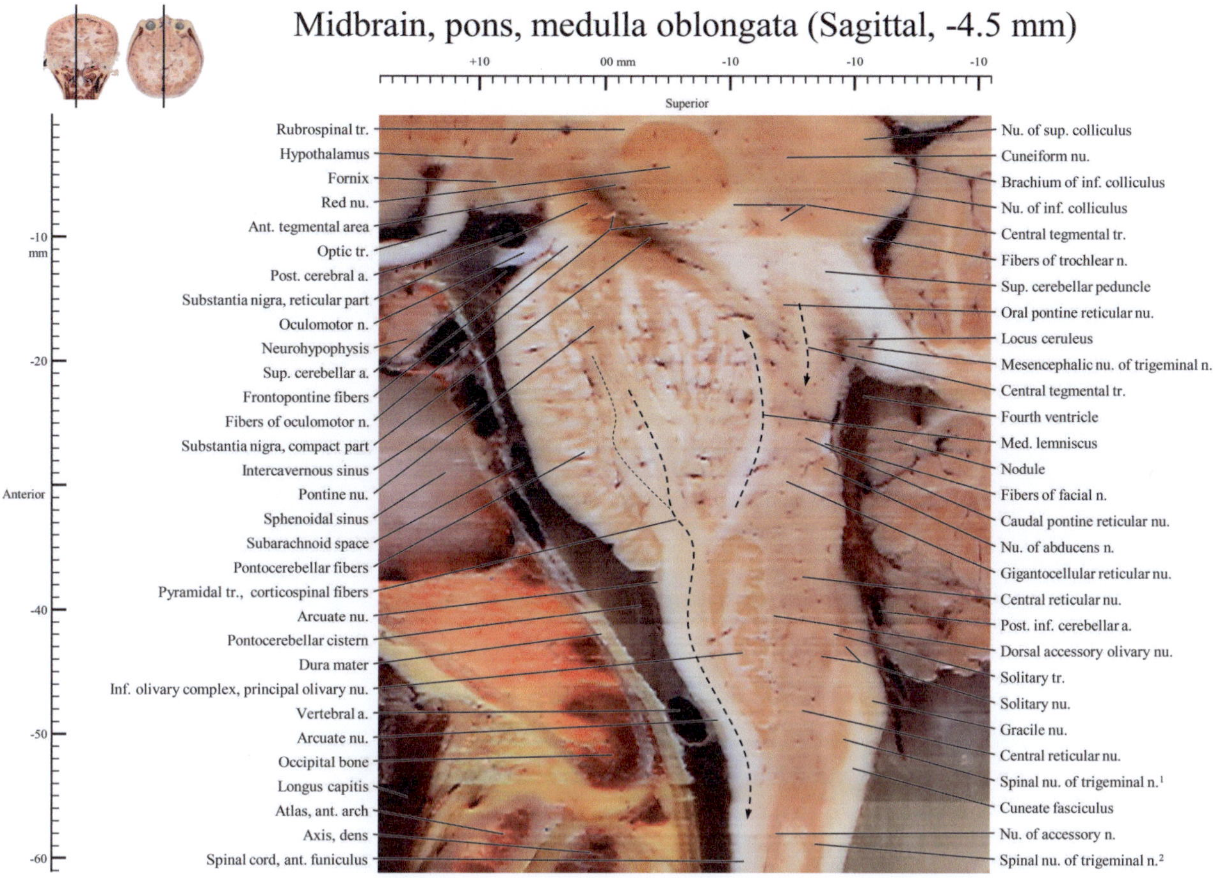

Midbrain, pons, medulla oblongata (Sagittal, -5.4 mm)

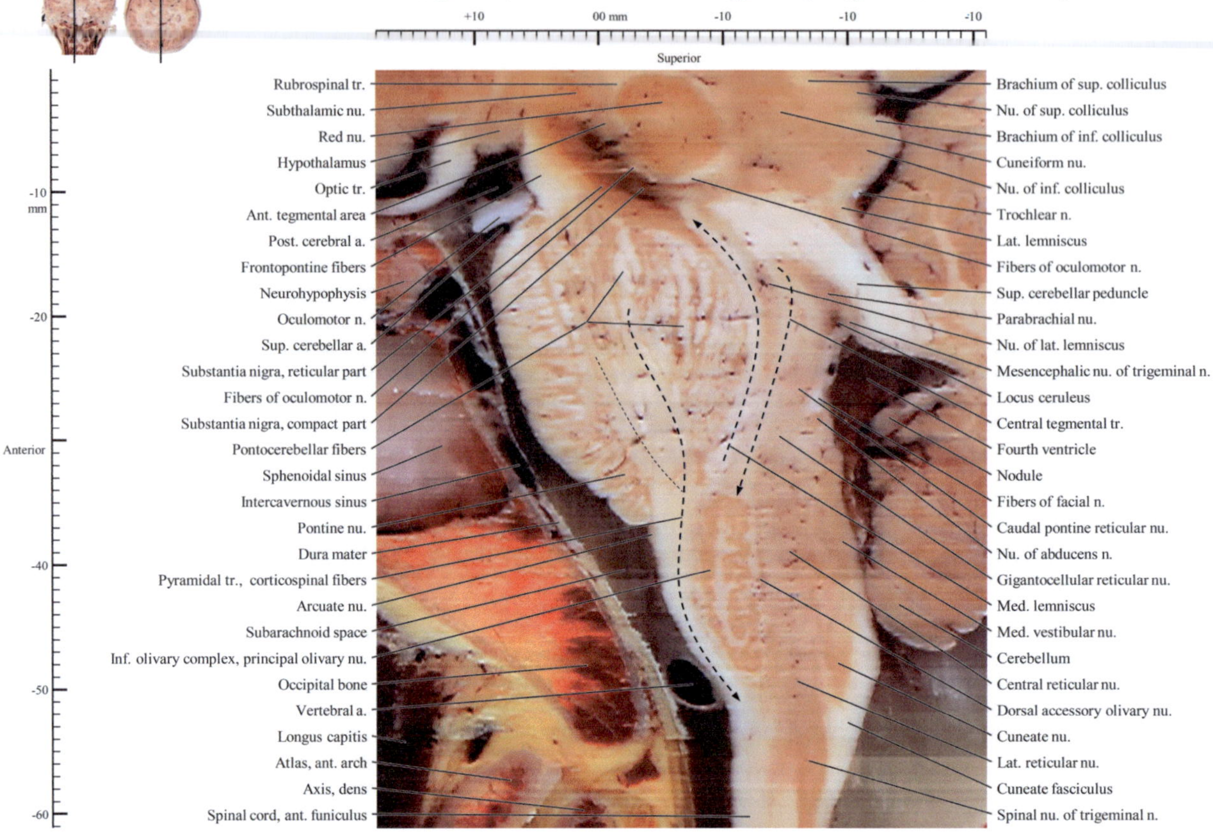

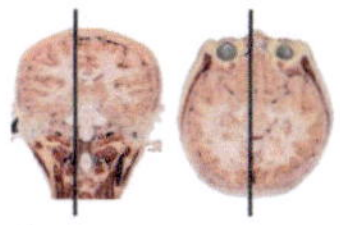

Midbrain, pons, medulla oblongata (Sagittal, -6.3 mm)

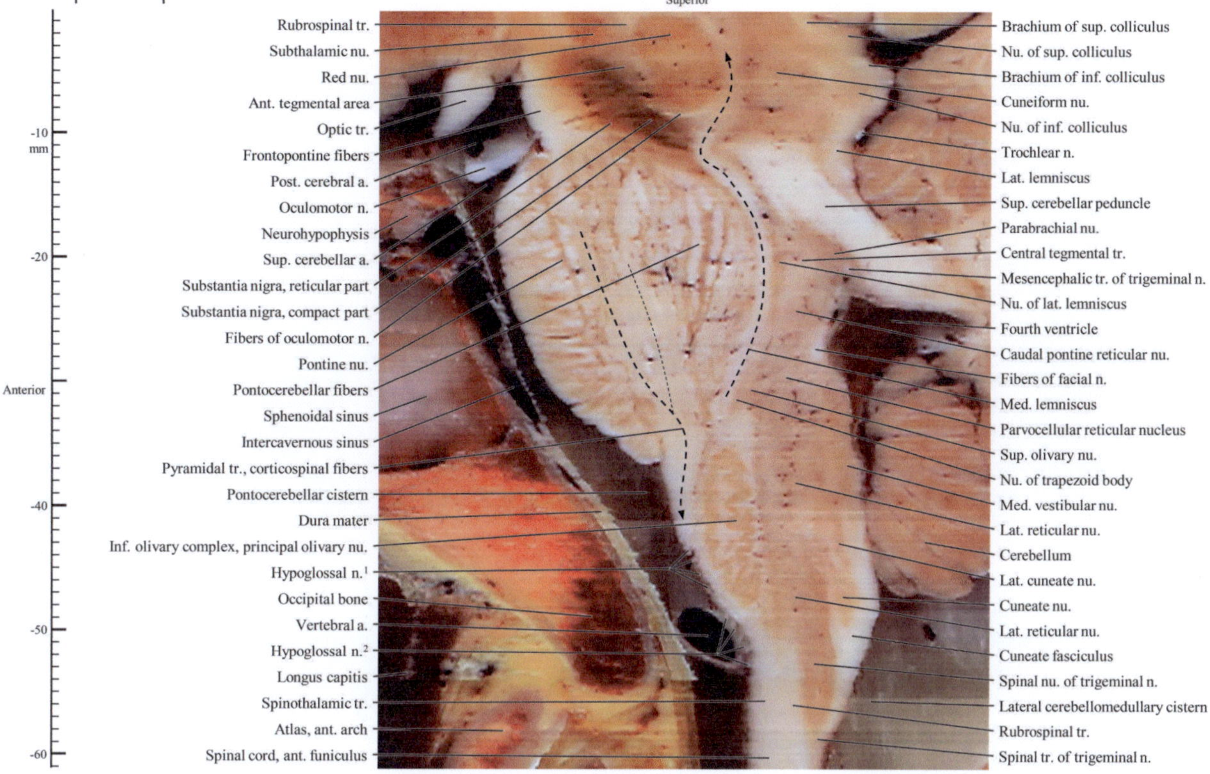

Midbrain, pons, medulla oblongata (Sagittal, -7.2 mm)

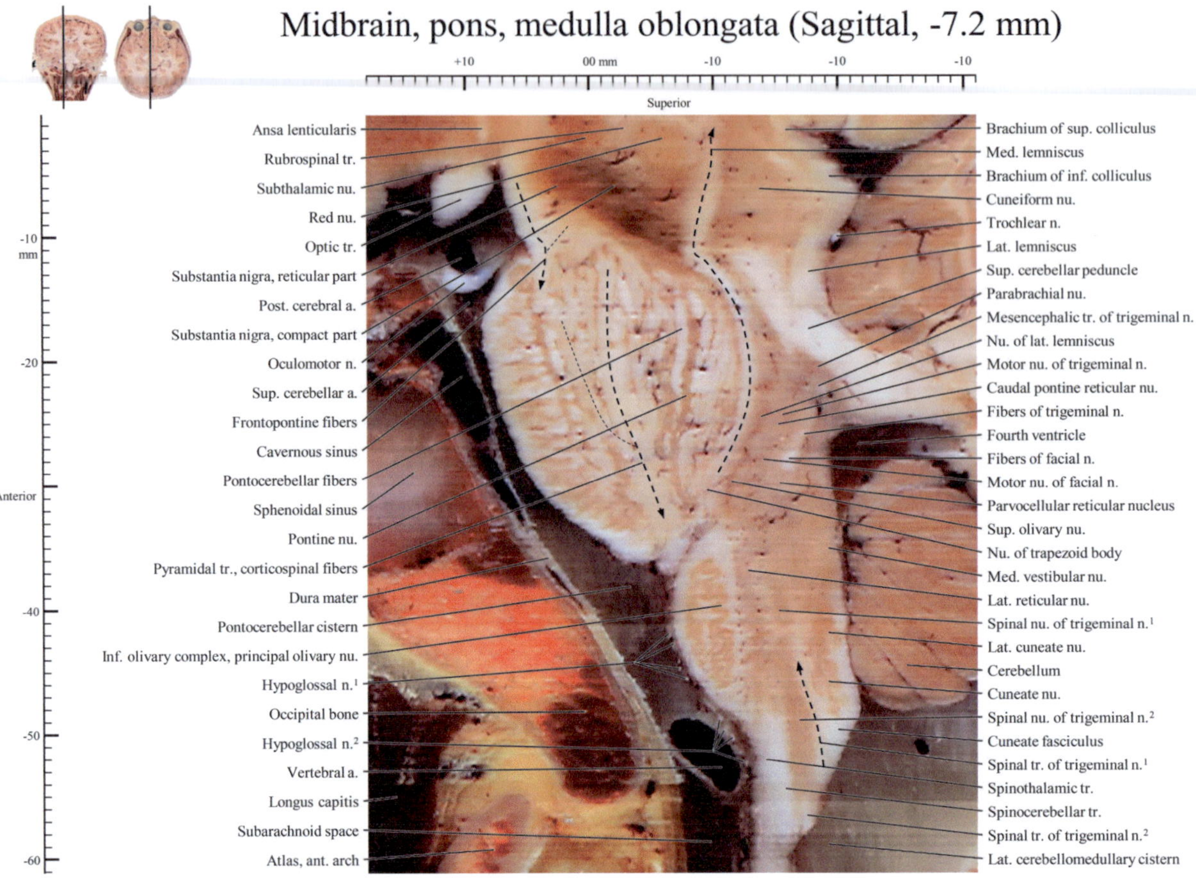

Midbrain, pons, medulla oblongata (Sagittal, -8.1 mm)

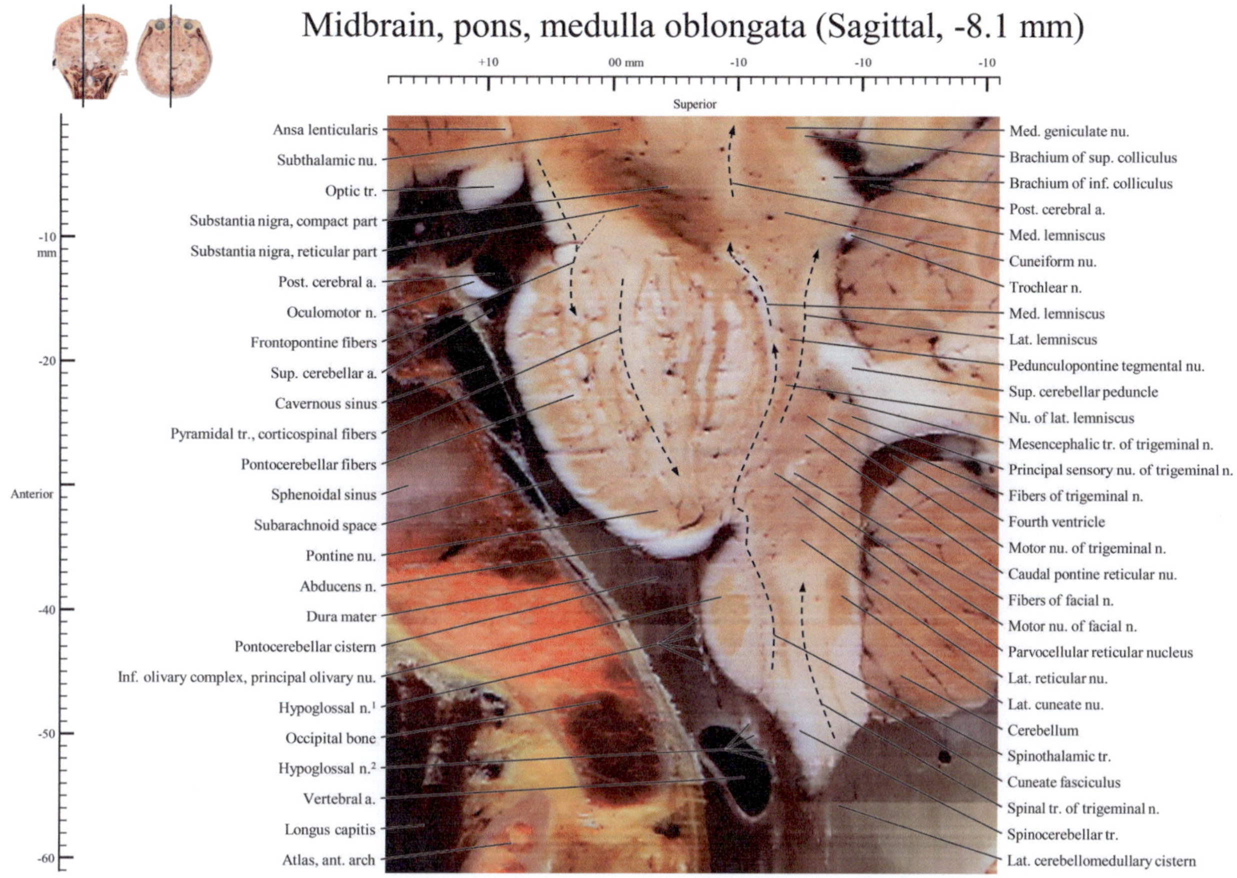

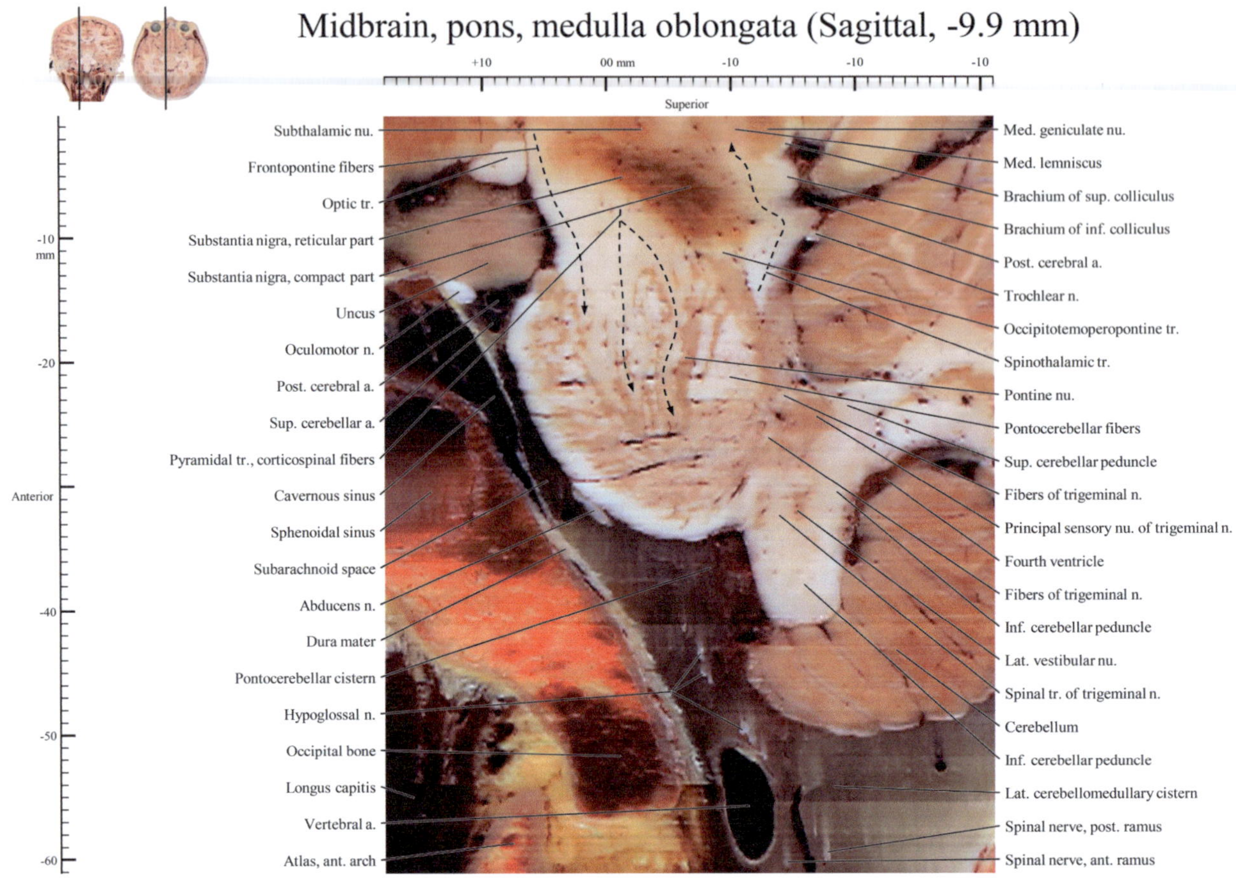
Midbrain, pons, medulla oblongata (Sagittal, -9.9 mm)
+10
00 mm
-10
-10
-10
Superior
-10 mm
-20
Anterior
-40
-50
-60
Subthalamic nu.
Frontopontine fibers
Optic tr.
Substantia nigra, reticular part
Substantia nigra, compact part
Uncus
Oculomotor n.
Post. cerebral a.
Sup. cerebellar a.
Pyramidal tr., corticospinal fibers
Cavernous sinus
Sphenoidal sinus
Subarachnoid space
Abducens n.
Dura mater
Pontocerebellar cistern
Hypoglossal n.
Occipital bone
Longus capitis
Vertebral a.
Atlas, ant. arch
Med. geniculate nu.
Med. lemniscus
Brachium of sup. colliculus
Brachium of inf. colliculus
Post. cerebral a.
Trochlear n.
Occipitotemoperopontine tr.
Spinothalamic tr.
Pontine nu.
Pontocerebellar fibers
Sup. cerebellar peduncle
Fibers of trigeminal n.
Principal sensory nu. of trigeminal n.
Fourth ventricle
Fibers of trigeminal n.
Inf. cerebellar peduncle
Lat. vestibular nu.
Spinal tr. of trigeminal n.
Cerebellum
Inf. cerebellar peduncle
Lat. cerebellomedullary cistern
Spinal nerve, post. ramus
Spinal nerve, ant. ramus

Midbrain, pons (Sagittal, -10.8 mm)

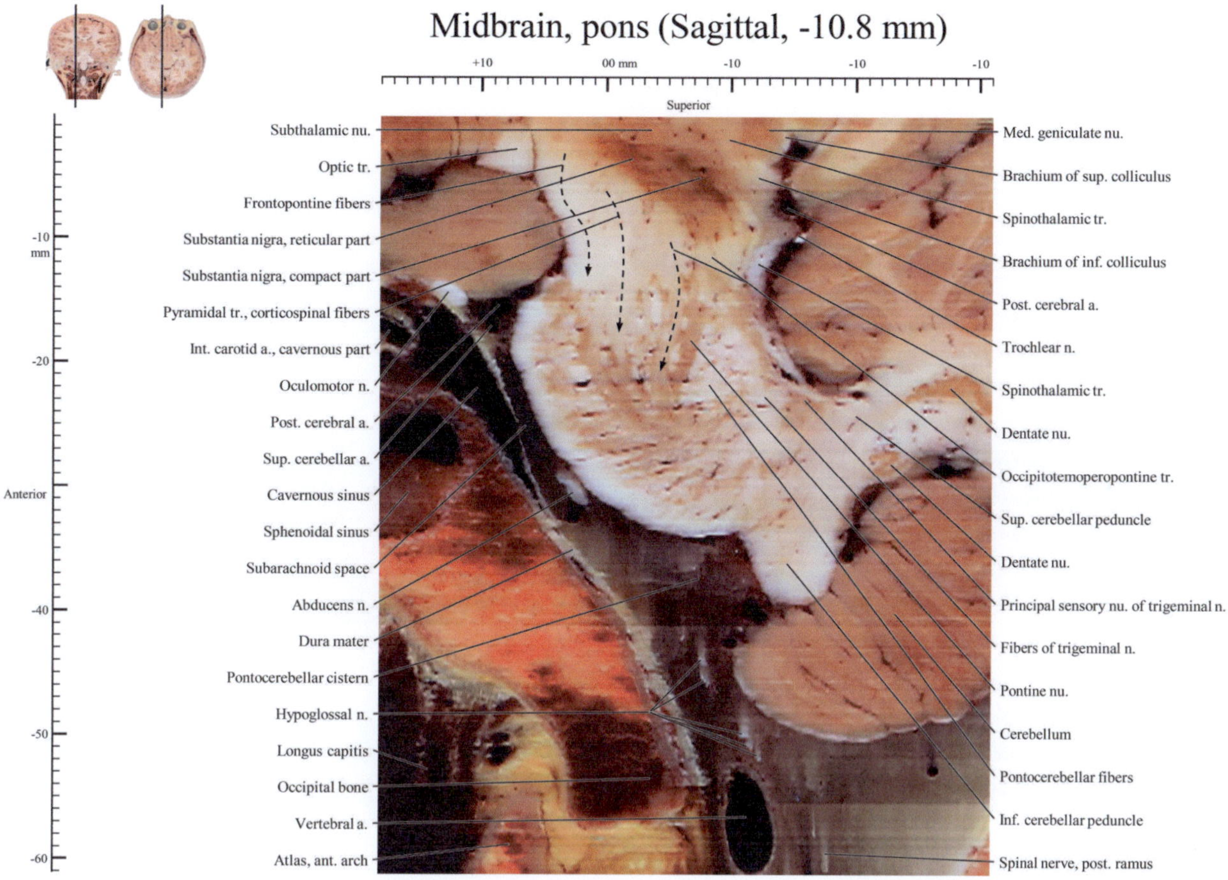

Midbrain, pons (Sagittal, -11.7 mm)

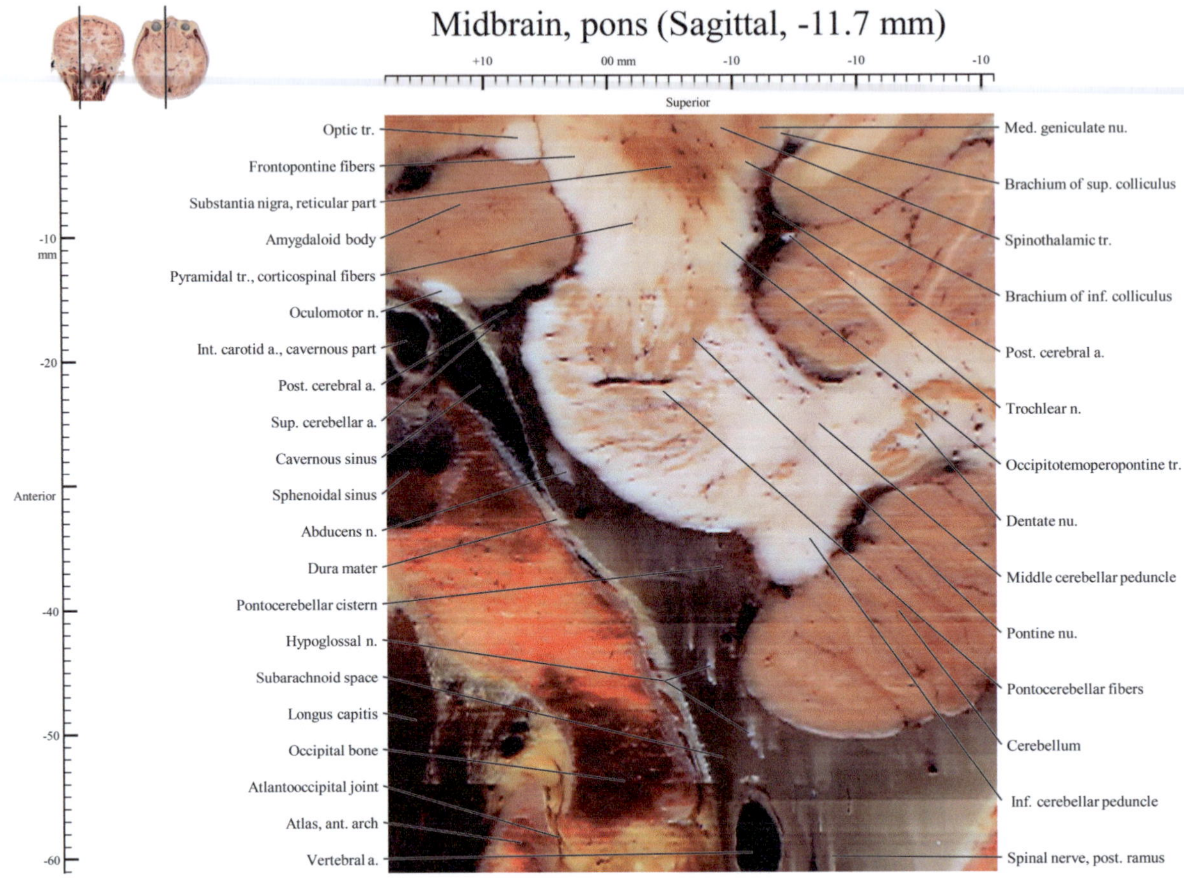

Midbrain, pons (Sagittal, -12.6 mm)

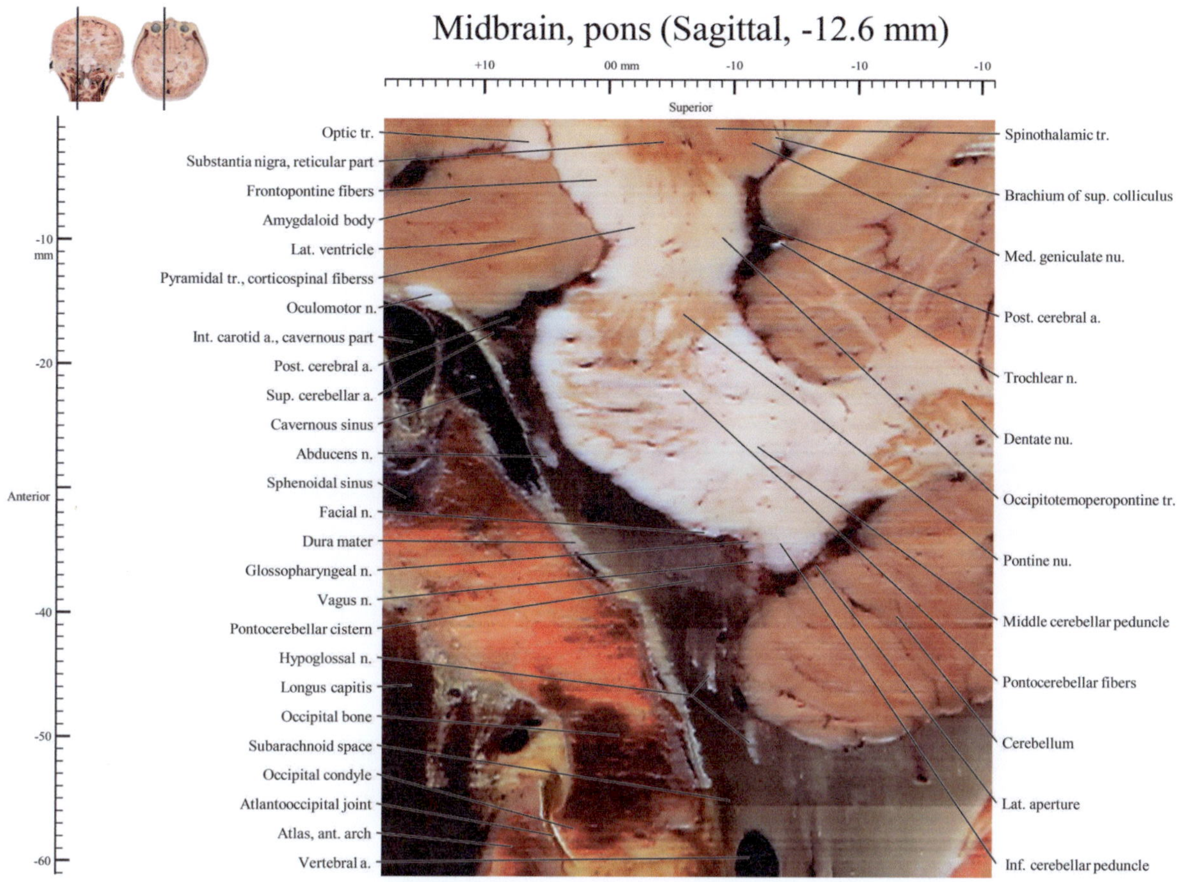

Midbrain, pons (Sagittal, -13.5 mm)

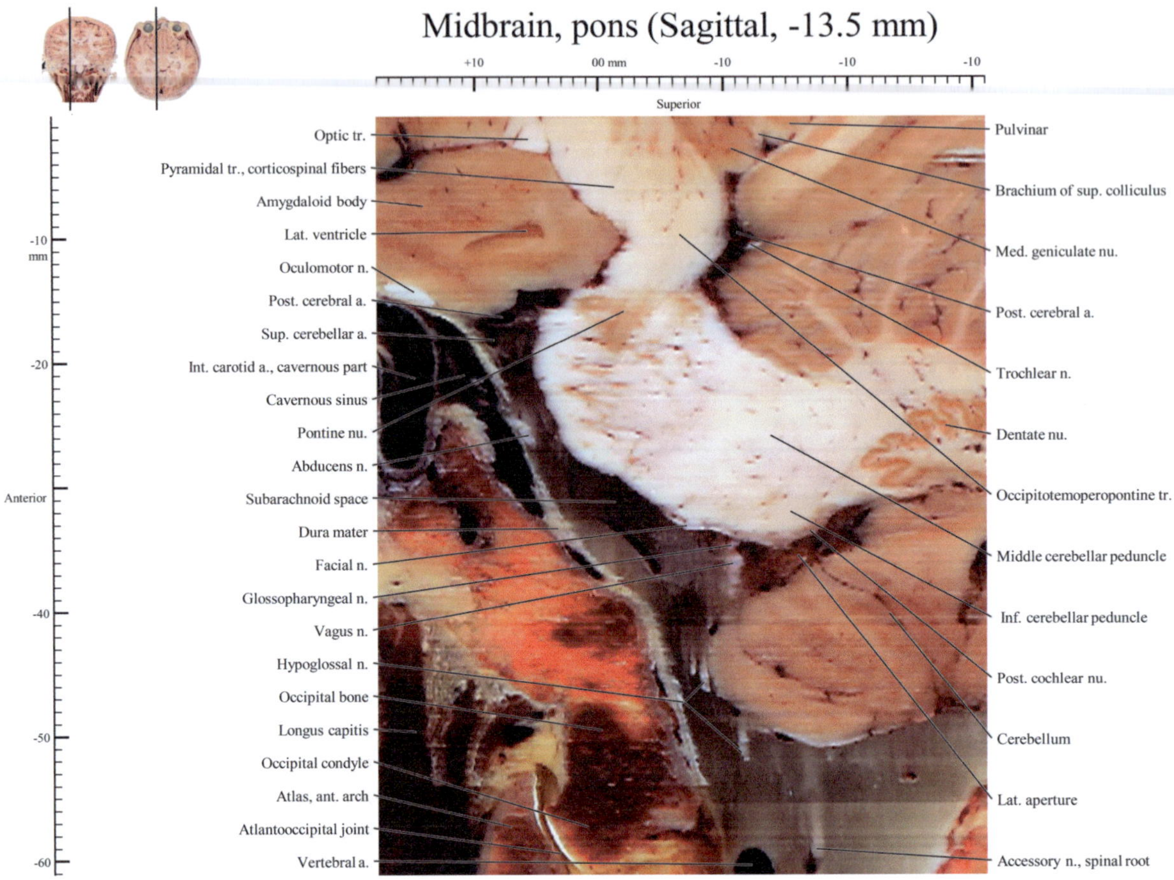

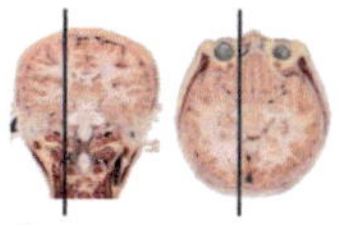

Midbrain, pons (Sagittal, -14.4 mm)

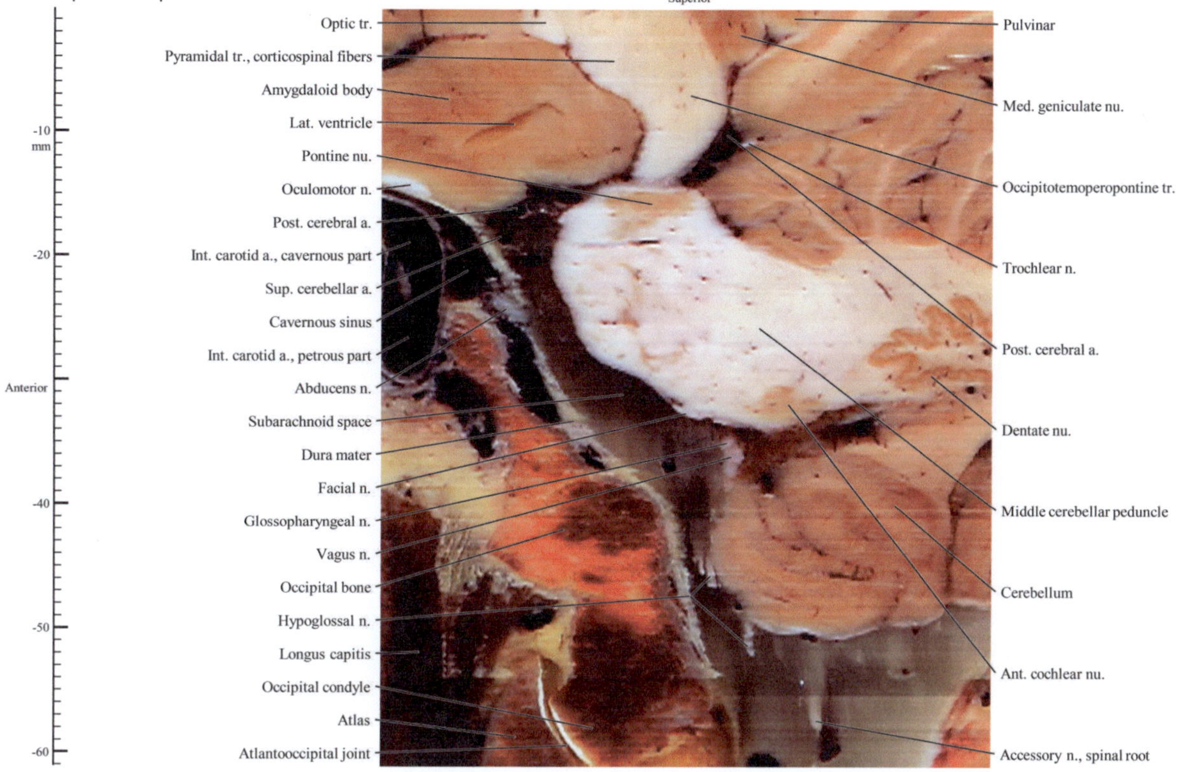

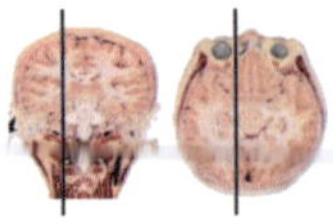

Midbrain, pons (Sagittal, -15.3 mm)

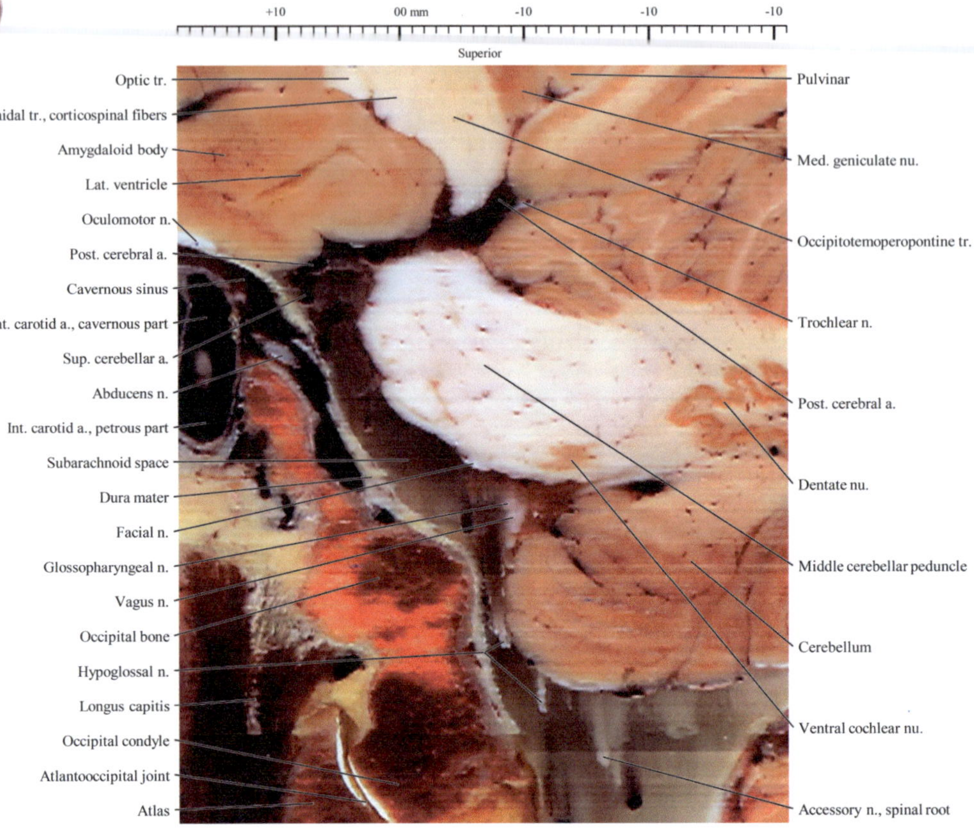

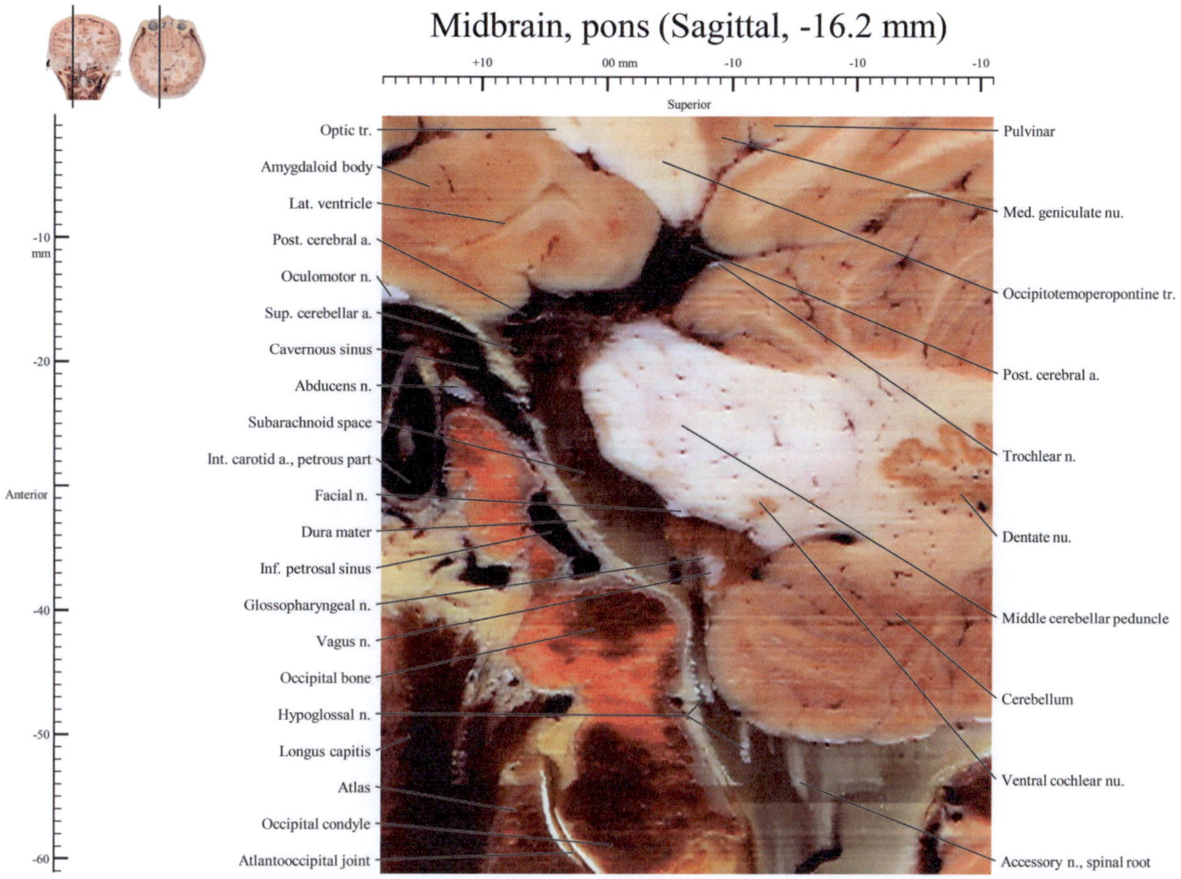
Midbrain, pons (Sagittal, -16.2 mm)
+10
00 mm
-10
-10
-10
Superior
-10 mm
-20
Anterior
-40
-50
-60
Optic tr.
Amygdaloid body
Lat. ventricle
Post. cerebral a.
Oculomotor n.
Sup. cerebellar a.
Cavernous sinus
Abducens n.
Subarachnoid space
Int. carotid a., petrous part
Facial n.
Dura mater
Inf. petrosal sinus
Glossopharyngeal n.
Vagus n.
Occipital bone
Hypoglossal n.
Longus capitis
Atlas
Occipital condyle
Atlantooccipital joint
Pulvinar
Med. geniculate nu.
Occipitotemoperopontine tr.
Post. cerebral a.
Trochlear n.
Dentate nu.
Middle cerebellar peduncle
Cerebellum
Ventral cochlear nu.
Accessory n., spinal root

Midbrain, pons (Sagittal, -17.1 mm)

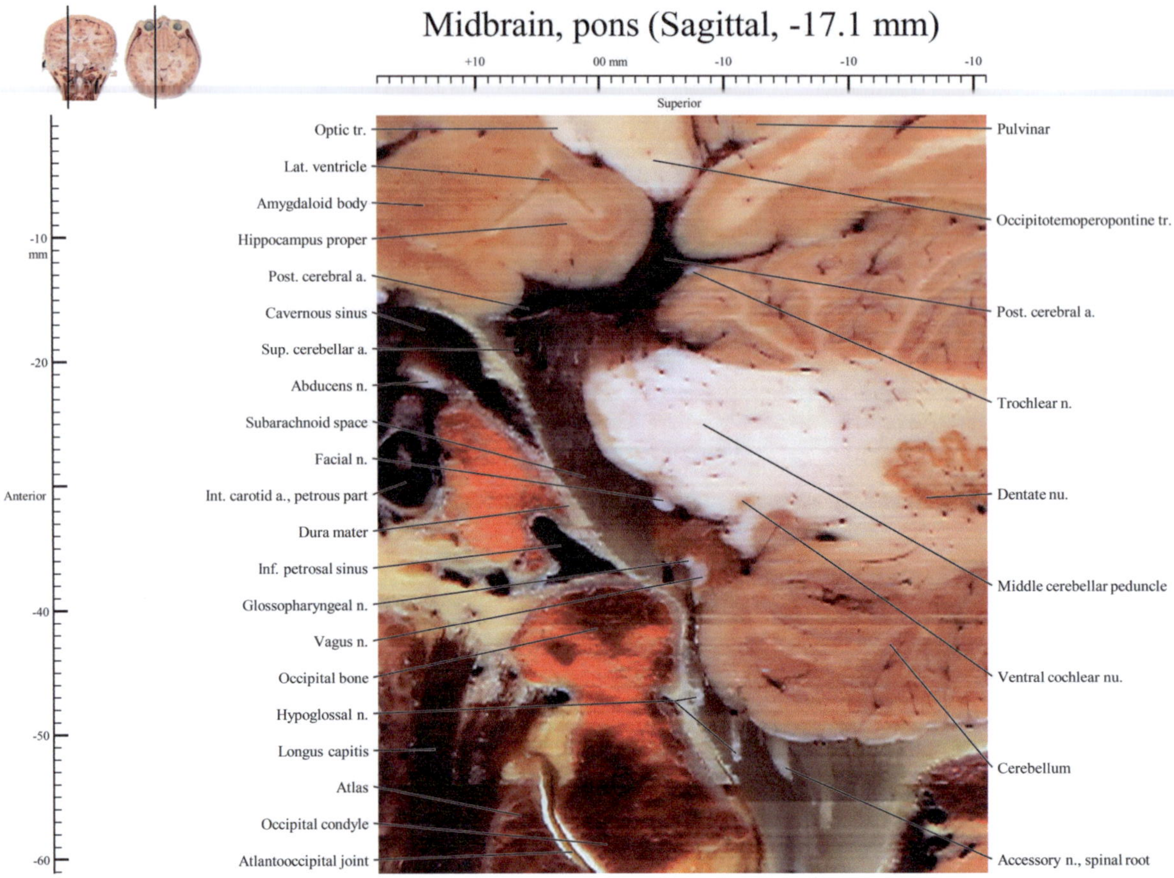

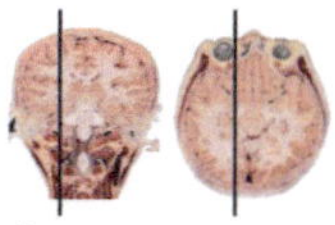

Pons (Sagittal, -18.0 mm)

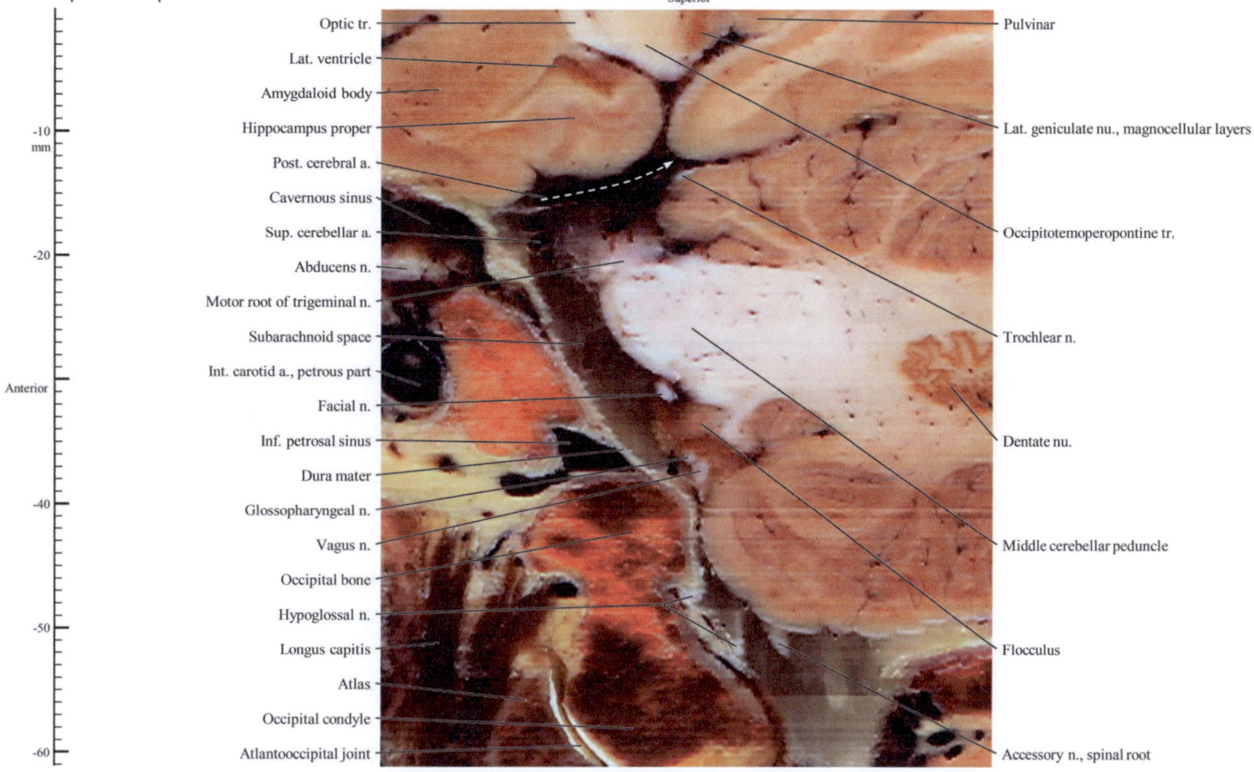

MIX
Papier aus verantwortungsvollen Quellen
Paper from responsible sources
FSC® C105338
www.fsc.org

If you have any concerns about our products,
you can contact us on
ProductSafety@springernature.com

In case Publisher is established outside the EU,
the EU authorized representative is:
Springer Nature Customer Service Center GmbH
Europaplatz 3, 69115 Heidelberg, Germany

Printed by Libri Plureos GmbH
in Hamburg, Germany